DYNAMIC ESTIMATION AND CONTROL OF POWER SYSTEMS

DYNAMIC ESTIMATION AND CONTROL OF POWER SYSTEMS

ABHINAV KUMAR SINGH
BIKASH C. PAL

ELSEVIER
ACADEMIC PRESS
An imprint of Elsevier

Academic Press is an imprint of Elsevier
125 London Wall, London EC2Y 5AS, United Kingdom
525 B Street, Suite 1650, San Diego, CA 92101, United States
50 Hampshire Street, 5th Floor, Cambridge, MA 02139, United States
The Boulevard, Langford Lane, Kidlington, Oxford OX5 1GB, United Kingdom

Library of Congress Cataloging-in-Publication Data
A catalog record for this book is available from the Library of Congress

British Library Cataloguing-in-Publication Data
A catalogue record for this book is available from the British Library

ISBN: 978-0-12-814005-5

Publisher: Jonathan Simpson
Acquisition Editor: Lisa Reading
Editorial Project Manager: Sabrina Webber / Michelle Fisher
Production Project Manager: Sruthi Satheesh
Designer: Greg Harris

Typeset by VTeX

Dedicated to our families
and our students

CONTENTS

ABOUT THE AUTHORS

Dr. Abhinav Kumar Singh is a lecturer at the University of Lincoln and a visiting researcher at Imperial College London, UK. He received his Bachelor of Technology (BTech) degree from Indian Institute of Technology Delhi, India, and his PhD from Imperial College London, in Aug. 2010 and Jan. 2015, respectively, both in electrical engineering. His research has focused on the application of control systems, signal processing, and communication systems for the estimation and control of power system dynamics. He has made novel contributions to the field, which have been recognized through various awards and recognitions, including the prestigious EPSRC Doctoral Prize Fellowship and the IEEE Power and Energy Society (PES) Working Group Recognition Award. He has entered into collaborations with power companies, such as Brush Electricals and ABB, for industrial implementation of his research. He is a Member of IEEE and has contributed to two Task Force reports, as well as chaired sessions and presented tutorials in IEEE PES General Meetings. (asingh@lincoln.ac.uk, a.singh11@alumni.imperial.ac.uk)

Prof. Bikash C. Pal is a professor of power systems at Imperial College London. He completed his BEng (1990) at Jadavpur University, India, his MEng (1992) at Indian Institute of Science, Bangalore, India, and his PhD (1999) at Imperial College London. He is active in research in power system stability, control, and computation. He has coauthored three books and two award winning IEEE Task Force/Working Group reports. He chairs an IEEE Working Group in state estimation for power distribution applications. As an IEEE distinguished lecturer, he has offered special lectures and tutorials in power system control and computation in more than 10 countries. He is consultant to power companies such as GE, ALSTOM, and National Grid. He was editor-in-chief of IEEE Transactions on Sustainable Energy (2012–2017), IET Generation, and Transmission & Distribution (2005–2012), and he is the series editor of Elsevier series on Sustainable Energies. He is a fellow of IEEE for his contribution to power system stability and control. (b.pal@imperial.ac.uk)

PREFACE

The purpose of this book is to introduce engineers and researchers of power systems to techniques for signal processing-based identification of the dynamic states and parameters of a power system and the use of the estimates for state space-based optimal control of the system, which are together known as dynamic estimation and control methods.

In order to meet the ever increasing demand for electrical power, modern energy grids are operating increasingly close to their security limits and are becoming more and more complex due to the increased integration of distributed sources of energy. Controlling such a network using traditional tools is becoming increasingly difficult and, hence, there is a shift in focus from static to dynamic monitoring and control of power systems. For controlling the system dynamically and adaptively, the operating state of the whole system needs to be identified in real-time, with estimation update rates in time scales of 10 ms. This fast estimation of the operating state is known as dynamic estimation, and the control methods based on dynamic estimation are referred to as dynamic control methods. The fast update rates of dynamic estimation ensure that the dynamic control methods remain valid for any operating condition and for any dynamic change in the system. Dynamic estimation and control is a fast growing and widely researched field of study, and it lays the foundation for a new generation of control methods which can dynamically, adaptively, and automatically stabilize power systems.

While disparate sources of research papers are available on this field, a single source containing a logically connected text explaining various concepts in a step-by-step manner, beginning from the fundamentals and building up to the most recent developments of the field, was missing. This book aims to fill this gap, which has been the main motivation for writing it. Hence, we have written this book to serve the twin goals of providing a consolidated source of thorough information for engineers and researchers working on dynamic estimation and control of modern grids and building the foundation of postgraduate students and researchers undertaking research in this area.

The book presents both centralized and decentralized, as well as linear and nonlinear theories for the dynamic estimation and control of power systems, and it is broadly based on our work in this area [128–133]. The applicability of all the presented methods has been demonstrated on a model 16-machine 68-bus benchmark system. The widely used MATLAB-Simulink software has been used as the software environment for coding and simulation.

In addition to explaining the concepts, significant focus has also been laid on mathematical proofs, the understanding of which requires readers to possess knowledge of undergraduate-level maths, in particular, the topics of linear algebra, differential equations, statistics, and probability. In general, one may choose to first understand the outline of the main ideas of a topic and leave the mathematics for later, if they desire a deeper technical understanding.

The book is self-contained and the fundamentals of power system modeling, simulation, and dynamic analysis, which are the building blocks for any estimation and control methodology, are explained in a concise but comprehensive and logical manner. Readers sufficiently well-versed with basics of power system estimation and control may skip the first two chapters and directly pick and choose the chapters of their interest, which in themselves present a complete discussion of their respective topics. Those who are new to the area of dynamic estimation and control, or to power systems in general, are advised to pursue a sequential reading through the text.

We hope that the research methods presented in the book will not only serve to create a clear understanding of the tools needed for estimating and controlling the dynamics of power systems in real-time, but will also encourage the readers to further explore and develop new and improved approaches in their own research.

We would like to express our most sincere gratitude to the members of Control and Power Group of Imperial College London, who spent time discussing our research with us on numerous occasions and provided their invaluable feedback, from which we have greatly benefited. A special note of thanks goes to Dr. Ravindra Singh for our earliest discussions on this field. Thanks also to the members of School of Engineering of University of Lincoln and the very helpful staff of Elsevier, especially, Ms. Sabrina, for guiding us throughout the process of writing. A very special thanks to Prof. Nilanjan Senroy, who has been a great inspiration for this work.

Above all, we are indebted to our families for always being there for us, through thick and thin. Without their continuous help and support, this book would not have been possible.

LIST OF FIGURES

LIST OF TABLES

LIST OF ABBREVIATIONS

AC	Alternating current
AGC	Automatic generation control
AVR	Automatic voltage regulator
CT	Current transformer
DAC	Discrete to analogue converter
DAE	Differential and algebraic equation
DC	Direct current
DFT	Discrete-time Fourier transform
DSE	Dynamic state estimation
DSP	Digital signal processing
EKF	Extended Kalman filter
ELQR	Extended linear quadratic regulator
EMS	Energy management system
FACTS	Flexible AC transmission system
GPS	Global positioning system
JLS	Jump linear system
LMI	Linear matrix inequality
LQG	Linear quadratic Gaussian
LQR	Linear quadratic regulator
LTI	Linear time-invariant
MPDP	Marginal packet delivery probability
NCPS	Networked controlled power system
NCS	Networked control system
NETS	New England test system
NYPS	New York power system
PDP	Packet delivery probability
PMU	Phasor measurement unit
POD	Power oscillation damping
PSS	Power system stabilizer
PT	Potential transformer (same as VT)
RTU	Remote terminal unit
SCADA	Supervisory control and data acquisition
SD	Standard deviation
SMIB	Single machine infinite bus
SSSC	Static synchronous series compensator
STATCOM	Static synchronous compensator
SVC	Static VAR compensator
TCP	Transmission control protocol
TCSC	Thyristor-controlled series capacitor
UDP	User datagram protocol
UKF	Unscented Kalman filter

WACS	Wide-area control system
WAMS	Wide-area measurement system
ZOH	Zero-order hold
VT	Voltage transformer (same as PT)

LIST OF SYMBOLS

$\mathbf{0}_{a\times b}$	a zero matrix of size $(a \times b)$
$\boldsymbol{\alpha}$	diagonal binary matrix representing input packet dropout
α	difference of rotor angle and stator voltage phase in rad
α^c	cth diagonal element of $\boldsymbol{\alpha}$
$\boldsymbol{\beta}$	diagonal binary matrix representing output packet dropout
β^c	cth diagonal element of $\boldsymbol{\beta}$
$\boldsymbol{\gamma}^-$	a predicted-measurement sigma point
δ	rotor angle in rad
ζ	damping ratio of an electromechanical mode
θ	stator voltage phase (or phase of a signal) in rad
θ_w	associated noise in the measured value of θ in rad
θ_y	measured value of θ in rad
λ_0	absolute bounding value of λ_y
λ_y	normalized innovation ratio for the measurement y
σ_η	standard deviation of η, for $\eta = V_w, \theta_w, I_w, \phi_w$, and f_w
$\boldsymbol{\phi}$	vector field of state mapping in normal form transformation
Φ	phase lead compensation required in degrees
ϕ	stator current phase in rad
ϕ_w	noise in the measured value of the stator current phase in rad
ϕ_y	measured value of the stator current phase in rad
$\boldsymbol{\chi}$	a sigma point
$\boldsymbol{\chi}^-$	a predicted-state sigma point
Ψ_{1d}	subtransient electromotive force (emf) due to d axis damper coils in p.u.
Ψ_{2q}	subtransient emf due to q axis damper coils in p.u.
ω	rotor speed in p.u. or in rad/s (depending on the context)
ω_b	base value of the rotor speed in rad/s
$\boldsymbol{A}$	state space matrix of system states
A_x, B_x	excitation system exciter saturation constants in p.u.
$\arg\{C\}$	angle of a complex number C in rad
$\boldsymbol{B}$	depending on the context, denotes state space matrix of inputs or denotes imaginary part of the bus admittance matrix in p.u.
$\boldsymbol{B}'$	state space matrix of system pseudoinputs
$\boldsymbol{C}$	state space matrix of system measurements
c	number of lead–lag stages
$\boldsymbol{D}(\boldsymbol{x})$	the vector $[L_f^{r_1} h_1(\boldsymbol{x})\ L_f^{r_2} h_2(\boldsymbol{x}) \ldots L_f^{r_m} h_m(\boldsymbol{x})]^T$
D	rotor damping constant in p.u.
$\text{diag}\{\boldsymbol{D}\}$	denotes a diagonal matrix of the elements in a vector $\boldsymbol{D}$
$\mathbb{E}[\boldsymbol{D}]$	denotes the expected value of a random variable $\boldsymbol{D}$
E_{fd}	field excitation voltage in p.u.
E_{fdmax}	upper-limit value of E_{fd} in p.u.
E_{fdmin}	lower-limit value of E_{fd} in p.u.

E'_d transient emf due to flux in the q-axis damper coil in p.u.
E'_{dc} state of the dummy rotor coil in p.u.
E'_q transient emf due to field flux linkages in p.u.
$\boldsymbol{F}$ depending on the context, denotes state feedback gain in the LQR and ELQR solutions, or denotes a vector of machine state functions
$\boldsymbol{f}$ vector of system state functions, $[f_1\ f_2 \ldots f_n]^T$
f frequency in p.u.
f_0 base value of frequency in Hz
f_s sampling frequency for interpolated-DFT method in Hz
$\mathbf{G}$ depending on context, denotes the real part of the bus admittance matrix in p.u., or denotes the feedback gain corresponding to $\boldsymbol{u}$ in the ELQR solution
$\mathbf{G}'$ feedback gain corresponding to $\boldsymbol{u}'$ in the ELQR solution
$\boldsymbol{g}$ a column vector of the system difference functions
$\boldsymbol{g}_i$ ith vector of input functions for $i = 1, 2, \ldots, M$
$\bar{\boldsymbol{g}}$ a column vector of the system differential functions
H generator inertia constant in s
$\boldsymbol{h}$ a column vector of output functions, $\boldsymbol{h} = [h_1\ h_2 \ldots h_M]^T$
$\bar{\boldsymbol{h}}$ a column vector of the system algebraic functions
h Hann window function
$\boldsymbol{I}_c$ denotes an identity matrix of size $(c \times c)$
I stator current magnitude in p.u.
I_{Di} D-axis component (in network's reference frame) of stator current in p.u.
I_{di} d-axis component (in generator's reference frame) of stator current in p.u.
I_{gi} generator current injection, $(I_{Qi} + jI_{Di}) + Y_{gi}V_{gi}$, in p.u.
I_{Qi} Q-axis component (in network's reference frame) of stator current in p.u.
I_{qi} q-axis component (in generator's reference frame) of stator current in p.u.
I_w noise in the measured stator current magnitude in p.u.
I_y measured stator current magnitude in p.u.
i refers to the ith generation unit or the ith bus in a power system
$\boldsymbol{J}$ quadratic cost for a discrete LTI system without pseudoinputs
$\boldsymbol{J}'$ quadratic cost for a discrete LTI system with pseudoinputs
j refers to $\sqrt{-1}$
$\boldsymbol{K}$ Kalman gain matrix
K_a Excitation system regulator gain in p.u.
K_c degree of compensation of a TCSC
K_{c-ref} control reference for the degree of compensation of a TCSC
K_{c-ss} control signal (change in degree of compensation) of a TCSC
K_{cmax} upper-limit value of K_c
K_{cmin} lower-limit value of K_c
K_{d1} the ratio $(X''_d - X_l)/(X'_d - X_l)$
K_{d2} the ratio $(X'_d - X''_d)/(X'_d - X_l)$
K_f excitation system feedback gain in p.u.
K_{pss} PSS gain in p.u.
K_{q1} the ratio $(X''_q - X_l)/(X'_q - X_l)$
K_{q2} the ratio $(X'_q - X''_q)/(X'_q - X_l)$
K_x excitation system exciter gain in p.u.
k refers to the kth time sample
$\bar{k}$ the $(k-1)$th time sample

$\boldsymbol{L}$	depending on the context, denotes the state feedback gain in stochastic LQR used in NCPS modeling, or denotes the Lie derivative
$\boldsymbol{l}$	refers to the left eigenvector of a given eigenvalue
l	refers to the lth sigma point
$\boldsymbol{M}$	positive definite matrix corresponding to $\boldsymbol{L}$
M	number of generation units in power system
m	number of elements in $\boldsymbol{x}$
N	depending on the context, denotes the number of buses in a power system, or the final time sample for reaching steady state in an LQR/ELQR solution
n	number of states or elements in $\boldsymbol{X}$
opt	denotes the optimal value of a variable
$\boldsymbol{P}$	positive definite matrix corresponding to $\boldsymbol{F}$
P	net active power injected at a bus in p.u.
P_{a-b}	denotes the active power flow in the line from bus a to bus b
P_G	active component of a power generation in p.u.
P_L	active component of a load in p.u.
$P_{s\tau}$	τth PSS state in p.u., for $\tau = 1, 2, 3$
$P'_{s\tau}$	τth PSS algebraic quantity in p.u., for $\tau = 1, 2, 3$
$\boldsymbol{P}_v$	covariance matrix of $\boldsymbol{v}$
$\boldsymbol{P}_w$	covariance matrix of $\boldsymbol{w}$
$\boldsymbol{P}_X$	estimated covariance matrix of $\boldsymbol{X}$
$\boldsymbol{P}_X^-$	estimated covariance matrix of $\boldsymbol{X}^-$
$\boldsymbol{P}_{Xy}^-$	estimated crosscorrelation between $\boldsymbol{X}^-$ and $\boldsymbol{y}^-$
$\boldsymbol{P}_x$	estimated covariance matrix of $\boldsymbol{x}$
$\boldsymbol{P}_{xz}$	estimated crosscorrelation between $\boldsymbol{x}$ and $\boldsymbol{z}$
$\boldsymbol{P}_x^-$	estimated covariance matrix of $\boldsymbol{x}^-$
$\boldsymbol{P}_y^-$	estimated covariance matrix of $\boldsymbol{y}^-$
$\boldsymbol{p}$	$(n-r) \times M$ matrix with (j, i) element $= L_{g_i}\phi_{j+r}(\boldsymbol{\phi}^{-1})$
p	number of elements in $\boldsymbol{u}$
p_u	PDP of an input channel
p_y	PDP of an output channel
$\boldsymbol{Q}$	state cost matrix
Q	net reactive power injected at a bus in p.u.
Q_G	reactive components of a power generation in p.u.
Q_L	reactive components of a load in p.u.
$\boldsymbol{q}$	$(n-r) \times 1$ vector with jth element $= L_f\phi_{j+r}(\boldsymbol{\phi}^{-1})$
q	number of elements in $\boldsymbol{y}$
$\boldsymbol{r}$	refers to the left eigenvector of a given eigenvalue
r	depending on the context, denotes number of elements in $\boldsymbol{u}'$, or denotes $r = \sum_{i=1}^{M} r_i$, where r_i is the relative degree of y_i
$\mathbb{R}$	denotes the set of real numbers
$\boldsymbol{R}$	input cost matrix
R	depending on the context, denotes residue, or denotes the reduced form of state space matrices, for example $\boldsymbol{A}_R$
$\boldsymbol{R}'$	pseudoinput cost matrix
R_a	armature resistance in p.u.
R_L	resistance of a line in p.u.

$\boldsymbol{S}$ matrix corresponding to $\mathbf{G}$ in the ELQR solution
$\boldsymbol{S}'$ matrix corresponding to $\mathbf{G}'$ in the ELQR solution
T denotes matrix transpose
T_0 system sampling period in s
$T_{\tau 1}$ τth stage lead time constants in s for $\tau = 1, 2$
$T_{\tau 2}$ τth stage lag time constants in s for $\tau = 1, 2$
T_a time constant of an excitation system's regulator in s
T_c time constant for the dummy rotor coil (usually 0.01) in s
T_{tcsc} time constant representing delay in TCSC firing sequence in s
T_e electrical torque input in p.u.
T_f excitation system feedback time constant in s
T_m mechanical torque input in p.u.
T_r time constant of excitation system's filter in s
T_w washout time constant in s
T_x time constant of excitation system exciter in s
T'_{d0} d-axis transient time constant in s
T'_{q0} q-axis transient time constant in s
T''_{d0} d-axis subtransient time constant in s
T''_{q0} q-axis subtransient time constant in s
t system time in s
$\boldsymbol{u}$ a column vector of the inputs to the system
$\boldsymbol{u}'$ a column vector of the pseudoinputs to the system
$\bar{\boldsymbol{u}}$ inputs to the system which have suffered packet dropout
$\boldsymbol{V}$ a column vector of the bus voltages, $V_i e^{j\theta_i}$, i=1, 2, . . . , N, in p.u.
V stator voltage magnitude in p.u.
V_a excitation system regulator voltage in p.u.
V_{Di} D-axis component (in network's reference frame) of stator voltage in p.u.
V_{di} d-axis component (in generator's reference frame) of stator voltage in p.u.
V_{gi} stator voltage, $(V_{Qi} + jV_{Di})$, in p.u.
V_{Qi} Q-axis component (in network's reference frame) of stator voltage in p.u.
V_{qi} q-axis component (in generator's reference frame) of stator voltage in p.u.
V_r excitation system filter voltage in p.u.
V_{ref} excitation system reference voltage in p.u.
V_{ss} PSS output voltage in p.u.
V_{ssmax} upper-limit value of V_{ss} in p.u.
V_{ssmin} lower-limit value of V_{ss} in p.u.
V_w associated noise in V in p.u.
V_y measured value of V in p.u.
$\boldsymbol{v}$ depending on context, denotes a column vector of process noise, or denotes a vector of inputs in normal form, $[v_1\ v_2 \ldots v_M]^T$
$\bar{\boldsymbol{v}}$ a column vector of process noise in continuous form
$\hat{\boldsymbol{v}}$ mean of a column vector of process noise
W DFT of Hann window function
$\boldsymbol{w}$ depending on context, denotes a column vector of measurement noise, or denotes an internal dynamics' state vector, $\boldsymbol{w} = [w_{r+1}\ w_{r+2} \ldots w_n]^T$
$\hat{\boldsymbol{w}}$ mean of a column vector of measurement noise
$\boldsymbol{X}$ augmented-state random variable

$\mathbf{X}^-$	predicted augmented-state random variable
X_C	capacitive reactance offered by a TCSC in p.u.
X_d	d-axis synchronous reactance in p.u.
X_L	reactance of a line in p.u.
X_l	armature leakage reactance in p.u.
X_q	q-axis synchronous reactance in p.u.
X_d'	d-axis transient reactance in p.u.
X_q'	q-axis transient reactance in p.u.
X_d''	d-axis subtransient reactance in p.u.
X_q''	q-axis subtransient reactance in p.u.
$\hat{\mathbf{X}}$	estimated mean of $\mathbf{X}$
$\hat{\mathbf{X}}^-$	estimated mean of $\mathbf{X}^-$
$\boldsymbol{x}$	column vector of the states
$\boldsymbol{x}^-$	predicted-state random variable
$\hat{\boldsymbol{x}}$	estimated mean of $\boldsymbol{x}$
$\hat{\boldsymbol{x}}^-$	estimated mean of $\boldsymbol{x}^-$
$\mathbf{Y}$	admittance matrix for network lines and generators, $\mathbf{Y} = \mathbf{Y}_N + \mathbf{Y}_G$, in p.u.
Y	a sinusoidal signal with harmonics and noise
$\mathbf{Y}_A$	augmented admittance matrix, $\mathbf{Y}_A = \mathbf{Y}_N + \mathbf{Y}_G + \mathbf{Y}_L$, in p.u.
$\mathbf{Y}_L$	load admittance matrix in p.u.
$\mathbf{Y}_G$	generator admittance matrix in p.u.
Y_g	generator admittance, $\frac{1}{(R_a + jX_d'')}$, in p.u.
Y_m	magnitude of Y's fundamental component in p.u.
$\mathbf{Y}_N$	bus admittance matrix for the power network in p.u.
$\boldsymbol{y}$	column vector of the observed measurements or outputs
$\boldsymbol{y}_a$	column vector of the algebraic variables of a system
$\bar{\boldsymbol{y}}$	observed measurements which have suffered packet dropout
$\boldsymbol{y}^-$	predicted-measurement random variable
$\hat{\boldsymbol{y}}^-$	estimated mean of $\boldsymbol{y}^-$
Z	DFT of the product of Y and h
$\mathbf{Z}_A$	augmented impedance matrix, $\mathbf{Z}_A = (\mathbf{Y}_A)^{-1}$, in p.u.
Z_a	armature impedance $(\sqrt{R_a{}^2 + X_d''^2})$ in p.u.
$\boldsymbol{z}$	depending on the context, denotes a column vector of noise in pseudoinputs, or denotes a vector of linearized states, $\boldsymbol{z} = [z_1^1 \ldots z_{r_1}^1 \ldots z_1^M \ldots z_{r_M}^M]^T$
$\hat{\boldsymbol{z}}$	mean of column vector of noise in pseudoinputs

CHAPTER 1

Introduction

Electrical power systems enjoy an enviable legacy of being one of oldest and most complex engineering systems ever built by mankind. Having reliable power systems is a key infrastructural requirement for the socio-economic development of the world. As power systems are considered to be the biggest and the most complex 'machines' ever built by mankind – the interconnected power system of North America is the largest human made machine on the earth – the control of these systems, so that they operate within their stability margins, is an equally complex and challenging task.

Stability of a power system is defined as "the ability of the system, for a given initial operating condition, to regain a state of operating equilibrium (or steady state of operation) after being subjected to a physical disturbance, with most system variables bounded so that practically the entire system remains intact" [1]. As alternating current (AC) of near-constant frequency is the most widely adopted standard for generation and delivery of power using synchronous machines, the most important criterion for stable operation is that all the synchronous machines in the system must remain in synchronism, or 'in-step'. This synchronism of generators in power systems is called *rotor angle stability*. More precisely, it refers to "the ability of synchronous machines of an interconnected power system to remain in synchronism after being subjected to a disturbance" [1].

A disturbance to the system can be a large one, such as a solid three-phase fault, or a small one, such as a small step change in system load. The ability of a power system to recover from a large disturbance is referred to as *transient stability*, while its ability to adequately dampen all the system oscillations after a small disturbance, or during steady-state operation, is referred to as *small signal stability*. Transient stability and small-signal stability together constitute rotor angle stability. Rotor angle stability is, thus, essential for power system operation and is achieved using corrective measures such as automatic generation control (AGC) as well as various other measures of protection and control [2].

Corrective measures of system protection and control should be taken in a timely manner once a disturbance has occurred, otherwise it can lead to system separation – a phenomenon in which the system divides into two

Dynamic Estimation and Control of Power Systems
https://doi.org/10.1016/B978-0-12-814005-5.00012-1

groups of machines and there is a loss of synchronism between the groups. Ultimately, this can lead to wide-scale blackouts and/or 'islanding' in the system, whereby the system breaks into many smaller subsystems, some of which – the so-called 'islands' – begin operating independently of the others, while the rest of the system enters into a state of blackout. To avoid such a crippling loss of synchronism, an example of a protective action in case of large disturbances is to clear a fault within its critical clearing time. An example of control action for both large and small disturbances is to damp the ensuing system oscillations using excitation systems of synchronous generators, power system stabilizers (PSSs), and/or power oscillation dampers (PODs) [2].

The growth of the power demand has been increasing at a steady pace, particularly in the developing countries, in contrast to the slow and incremental nature of the evolution of control technologies for power systems. The grid interconnections have increased manifold and there is an assimilation of more and varied (both centralized and decentralized) sources of energy into the grid. Deregulation has led to increased separation of power producers and consumers, and there is an increased demand for not only power but also for high-quality power. In order to meet these growing demands, power systems have not only grown larger and more complex than ever (mainly due to large scale interconnections and integration of renewable sources of energy), but they are operating increasingly close to their operating limits as well, as elaborated in [3]. The European Network of Transmission System Operators for Electricity (ENTSO-E)-interconnected system is an example of a stressed power system which is being operated more and more near its limits [3]. Small signal stability of such stressed systems is becoming more difficult to achieve using traditional schemes based on excitation systems of generators and PSSs. For instance, in some power blackout analyses the ineffectiveness of the control of small signal stability was identified as an important link to the inception of the events leading to system-wide blackouts [4,5].

It has been observed that under certain conditions, a small disturbance in a power system can initiate spontaneous oscillations in the power flows in the transmission lines. These oscillations grow in magnitude within few seconds if they are undamped or poorly damped. This can lead to a loss in synchronism of generators or a voltage collapse, ultimately resulting in system separation and blackouts. The power blackout of August 10, 1996 in the Western Electricity Co-ordination Council region is a famous example of blackouts caused by such oscillations [6,7]. The frequencies of these oscillations are in the range of 0.2 to 1.0 Hz, and as these oscillations are not local to a particular generator and involve two or more groups of generators (also known as areas), they are termed interarea oscillations [8,9]. The local control actions of excitation systems of generators and PSSs are insufficient to control interarea oscillations, and therefore more global control schemes are needed to achieve small signal stability in current power systems.

The abovementioned reasons have provided an impetus to the research, development, and investment in global control schemes for power systems in recent times. Phasor measurement units (PMUs) and flexible AC transmission system (FACTS) are starting to form the core of such a global control infrastructure for power systems. New techniques for dynamic state estimation (DSE) and dynamic control are emerging which can not only strengthen but potentially revolutionize the control infrastructure. The next section explores the state of the art and current research in power system estimation and control.

1.1 STATE OF THE ART

1.1.1 Energy management system

The energy management system (EMS) in a power system plays an important role in system operation and control [10]. EMS has a host of network computation functions, such as static state estimation (SSE), optimal power flow, and contingency analysis. In EMS, the estimates of the operating states of the system are updated in the time scale of a second or more through SSE. SSE is used for scheduling and dispatch of load and generation, but because of slow update rates, it is not suitable for real-time monitoring and control of time-critical dynamics in the system.

The supervisory control and data acquisition (SCADA) system forms the heart of EMS and performs data acquisition, updates of the system status through alarm processing, updates the user interface, and executes control actions [11]. Remote terminal units (RTUs) perform the role of sensors and actuators in SCADA systems. Different types of telemetering and communication protocols are used in SCADA systems (which vary with SCADA vendors), but the majority of them use serial communication based on the DNP3.0 protocol [12]; their update rates lie in the range of 2–10 seconds [13,14]. Although these rates are fast enough to provide the traditional functions performed by EMS, they are not enough to deliver time-critical measurements and control actions needed for dynamic estimation and control. Besides communication systems, there are several aspects of EMS/SCADA (such as metering, security, visualization, database, and control capabilities) that need to be upgraded to meet the requirements of today's power systems [15].

1.1.2 Phasor measurement units (PMUs)

An electrical quantity which has both phase and magnitude (for example bus voltage, line current, and line power) is called a phasor. A PMU is a device which can accurately measure a phasor. This is done by the time synchronization of all the PMUs in the power system to an absolute time reference provided by the global positioning system (GPS) [16,17]. PMUs are capable of providing sampling rates of over 600 Hz and time synchronization accuracy of ± 0.2 μs [18]. The speed and accuracy of measurements by PMUs have led to the development of several techniques and algorithms for dynamic state estimation and fast and reliable control methods.

1.1.3 Flexible AC transmission system (FACTS)

FACTS devices are static power-electronic devices installed in AC transmission networks to increase power transfer capability, stability, and controllability of the networks through series and/or shunt compensation [19]. These devices are also employed for congestion management and loss optimization. The static synchronous series compensator (SSSC) and thyristor-controlled series capacitor (TCSC) are some of the FACTS control devices which provide series compensation to reactance of the lines to which they are connected, while the static synchronous compensator (STATCOM) and static VAR compensator (SVC) (where VAR stands for volt–ampere reactive) are FACTS devices which provide shunt compensation to transmission lines. FACTS control devices also provide adequate damping of interarea oscillations by acting as actuators in robust control schemes and PMU-based wide-area control schemes [7,19].

1.1.4 Wide-area measurements and wide-area control

Wide area measurement system (WAMS) refers to a measurement system composed of strategically placed time-synchronized sensors (which are PMUs) which can monitor the current status of a critical area in real-time. The critical area can be the entire power system or a part of the system. Strategic locations for placing PMUs are decided in such a way that the number of locations is minimized while the critical area remains completely observable [20]. The measurements from the WAMS are utilized by the wide-area control system (WACS) to control the transient and oscillatory dynamics of system voltage and frequency [21]. A fast communication network which can operate at update rates of 10–20 Hz is crucial for the WAMS/WACS in order to deliver measurements from sensors to the control center and control signals from the control center to actuators (excitation system of generators, PSSs, and FACTS devices). As the communication requirements of WAMS/WACS are very high, at present WAMS/WACS have only been implemented in small-scale power systems. The WAMS/WACS implemented by Bonneville Power Administration for the wide area stability and voltage support of their power system is one such example [21]. The communication architecture for power systems needs to be revamped in order to implement WAMS/WACS on large-scale power systems [22–24].

1.1.5 Dynamic state estimation (DSE) and dynamic control

DSE, which refers to the estimation of state variables representing oscillatory dynamics of a power system, is also utilized for effective control of these dynamics besides the aforementioned techniques of robust control and wide-area control. With growing deployment of PMUs across the system, DSE algorithms have been proposed by several research groups for real-time estimation of dynamic states (typically machine rotor angle, speed, transient speed voltages, etc.) using Kalman filtering [25–31]. However, as many of these algorithms present a centralized approach to DSE, to implement them, a reliable and fast communication network is needed to bring system-wide measurements to a central location in real-time. Thus, a slow communication network used in EMS/SCADA (with update rates of 2–10 s) creates a bottleneck for both WACS and DSE. There is also a growing concern about the reliability and security of the current communication networks. As a solution to this problem, several new methods of DSE have also been proposed in recent literature [32–41], which are discussed in Section 1.3.

DSE forms an integral part of many dynamic control techniques proposed for today's power system. Algorithms based on real-time dynamic security assessment ([42,43]), model predictive control ([44,45]), and nonlinear and linear optimal control methods ([46–61]) are examples of such control techniques.

1.2 STATIC STATE ESTIMATION (SSE) VERSUS DYNAMIC STATE ESTIMATION (DSE)

The primary difference between SSE and DSE is the type of power system model used for estimation (that is, whether the power system model is static or dynamic) and the time scale of estimation. In SSE, the quasisteady state or static operating behavior of a power system is estimated through the estimation of voltage magnitudes and phases of all the nodes or buses in the network. This requires measurement update rates in the time scale of a few seconds. On the other hand, DSE estimates the underlying oscillatory dynamics of power system operation through the estimation of system states which describe these oscillatory dynamics, such as the rotational speed of a generator in the system. DSE requires update rates in the order of tens of milliseconds. Thus, SSE works with a 'static structure' of the system, which consists of network generator–load geometry and the various transmission lines, while DSE works with a "dynamic structure" of the system, in which the dynamics of the generators and other dynamic components in the network are modeled along with the static structure.

The utility of SSE is also widely different from that of DSE. SSE is required for load generation dispatch and for monitoring of bus voltage levels and line power levels of various buses and transmission lines in the system. DSE, however, is utilized for decentralized and/or wide-area control of system dynamics in order to ensure system stability and for monitoring the oscillations occurring in the system.

1.3 CHALLENGES TO POWER SYSTEM DYNAMIC ESTIMATION AND CONTROL

The majority of control and monitoring tools in present power systems are provided by EMSs and are based on steady-state system models, which cannot capture the dynamics of the power system very well. This limitation is primarily due to the dependency of EMSs on slow update rates of the SCADA systems. Due to slow update rates, the state estimates of the system are updated in a time scale of a few seconds, and most of the dynamic control schemes are local to a generator or a FACTS device and are based on locally available information and measurements. The main challenge in implementing dynamic estimation and global control schemes is the unavailability of a fast, cost effective, reliable, and secure communication network.

Packet-based communication is the most widely adopted communication technology today, on which even the highly complex "internet" is based. There is an option of using a packet-based communication network (instead of the slow and outdated communication technology used in SCADA) for DSE, WAMS/WACS, and dynamic control in power systems. But this option also poses a question regarding the stability of the overall system, as packet-based communication suffers from various problems, such as packet dropout, packet disordering, and time delays. There are also problems of multivendor PMU accuracy, data wrapping, etc. associated with WAMS technology.

Another option for implementing DSE, WAMS/WACS, and dynamic control in power systems is to implement them in a completely decentralized manner. This means that the complete knowledge of states and controllability of the oscillatory system dynamics are obtained at decentralized locations in the system using only local information and measurements at those locations. This option removes the necessity of a fast and reliable communication network for dynamic estimation and control. Owing to the practicability of this solution, research focus in the development of decentralized methods of DSE has increased significantly [32–41]. Decentralized control of power systems is also a developing area of research and several recent algorithms have been proposed ([47–57]). However, most of the current algorithms assume a simplistic model of power systems and use noisy (unfiltered) measurements for control implementation.

1.4 BOOK ORGANIZATION

The organization of the rest of the book is as follows. Chapters 2 and 3 present the preliminary background concepts of dynamic estimation and control. Chapter 2 is a review of the basics of power system modeling and simulation, which form the building blocks for developing estimation and control methodologies. In Chapter 3, a framework for centralized dynamic estimation and control using a packet switching-based networked control system (NCS) in power systems has been presented. Some practical limitations and problems associated with real-time centralized estimation and control using NCS are also discussed.

Chapters 4–6 are related to decentralized estimation. Chapter 4 presents the method for decentralization of dynamic estimation. This method is based on the concept of pseudoinputs in which some of the measurements are treated as inputs. Unscented Kalman filtering is then applied on the decentralized power system model for DSE using PMU measurements. Chapter 5 presents the theory of signal parameter estimation using the analogue signals obtained from instrument transformers, and Chapter 6 demonstrates how these signal parameters can be utilized for decentralized DSE without requiring time synchronization through GPS or any other source of synchronization.

Chapters 7–9 present methods for decentralized control of power systems. Various linear and nonlinear theories based on dynamic estimation are presented in Chapter 7 for the optimal control of a power system. Out of these theories, an extended linear quadratic regulator is described in Chapter 8 for decentralized control of each generation unit in a power system such that the whole power system is stabilized and all oscillations in the system are adequately damped. Similarly, in Chapter 9, a normal form-based nonlinear controller is used for ensuring both transient and small signal stability of a power system in an optimal manner. A description of the system used in case studies and some of the implementation details for MATLAB are presented in the appendices.

CHAPTER 2

Power System Modeling, Simulation, and Control Design

This chapter provides a basic conceptual guide to power system modeling, simulation, and control design, illustrating the ideas through application on the example test system presented in Appendix A. The simulation codes for each of the concepts discussed here have been made available online [134], with slightly modified parameters.

A power system is a large-scale complex network of components spread over a wide geographical region. Hence, building its physical prototype is prohibitive, and simulation has been the common practice for studying its operation and behavior. Simulation first requires a model to be developed; in order to analyze the dynamics and stability of a power system, its dynamic model is required. But power system dynamics are very complex, involving different time scales for different components in the system. Therefore, we begin the chapter with a discussion of the dynamics of such power system components which are primarily responsible for influencing the overall system dynamics and build their dynamic models. Next, the overall dynamic behavior of the power system is analyzed through simulations of the integrated model. This chapter concludes with a discussion about the basics of the control design framework for designing an economic, practical, and efficient methodology for controlling the system's oscillatory dynamics.

2.1 POWER SYSTEM MODEL

A power system is an interconnected system consisting of various components, such as the generators, their excitation systems and control mechanisms, power system stabilizers (PSSs), flexible alternating current (AC) transmission system (FACTS) control devices (such as a thyristor-controlled series capacitor (TCSC) for series compensation and static VAR compensator (SVC) for shunt compensation), loads, and the transmission network [62,63]. Power electronics converter-interfaced generation from solar and wind sources are also increasingly being integrated to the system. Proper modeling of each of these components is fundamental to the study of the dynamic behavior of a power system.

Dynamic Estimation and Control of Power Systems
https://doi.org/10.1016/B978-0-12-814005-5.00013-3

There are two types of variables that govern power system dynamics: the electrical variables with fast dynamics of frequencies close to synchronous frequency (that is, 50 or 60 Hz), such as power flow in transmission lines and bus voltages, and the electromechanical variables with slow dynamics of a frequency less that 3 Hz, such as generator speed. The former are approximated through algebraic equations, while the slow electromechanical dynamics of the system are modeled using a set of nonlinear differential equations. This leads to a system of differential and algebraic equations (DAEs) ([62,63]). Note that the dynamic response of a power system to any disturbance in the system is nonlinear in nature. This can be easily appreciated as the power output of synchronous generators and power flow in the network are a product of nodal voltage magnitudes and sinusoids of the angles between voltages, which will be reflected in the DAEs. A brief description of power system components and the DAEs that govern them are presented below. Unless specified, all symbols used in the DAEs are as defined in the "List of Symbols".

2.1.1 Generating unit: a generator and its excitation system

Each generating unit of a power system consists of a synchronous generator and its excitation system. A synchronous generator has a stator connected to the system and a rotor connected to the prime mover. Mechanical power is converted to electrical power through electromechanical interactions at synchronous speed and transferred from the rotor to the stator.

The dynamic interactions between the rotor and the stator decide the modeling complexity of the generator. The recommended practice as per IEEE standard is to represent each generator in the system with four equivalent rotor coils. The very slow mechanical dynamics of the governors (or the prime movers) are ignored, and the mechanical torques to the generators are taken as constant inputs.

In a generator, mechanical energy is transformed to electrical energy only if the field windings of the generator are excited using excitation voltage and current. Hence, each generator in the system is provided with an excitation system to control the field excitation voltage of the generator. Excitation system models have been classified and standardized by IEEE [64]. Fig. 2.1 and Fig. 2.2 show the block diagrams of two of the most common excitation systems – the IEEE-ST1A static excitation system and the IEEE-DC1A direct current (DC) excitation system, respectively. The IEEE-DC1A model is used for field-controlled DC commutator exciters with continuously acting voltage regulators and generator terminal voltage

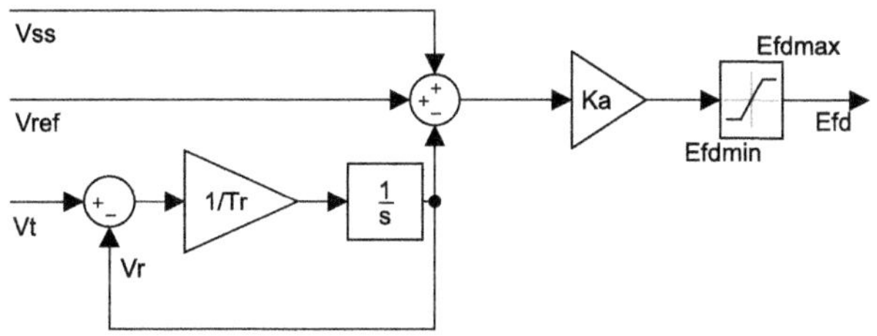

Figure 2.1 Block diagram of IEEE-ST1A static excitation system.

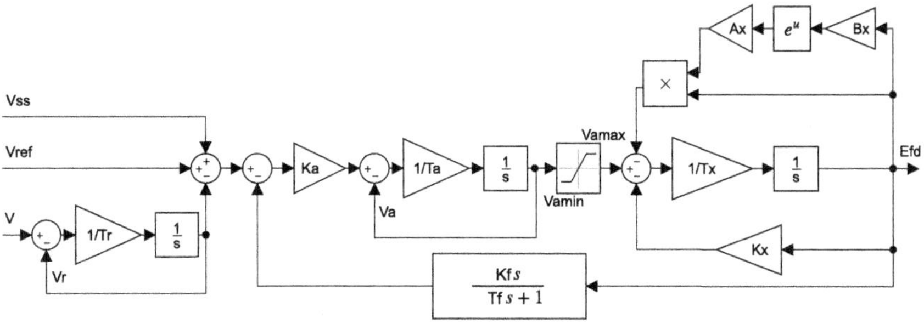

Figure 2.2 Block diagram of IEEE-DC1A DC excitation system.

as main input. The IEEE-ST1A model is used to represent an excitation system in which excitation power is supplied through a transformer from the generator terminals (or the unit's auxiliary bus) and is regulated by a controlled rectifier.

The DAEs of a generating unit in a power system (that is, the equations of a machine and its excitation system) are given as follows:

$$\dot{\boldsymbol{x}} = \boldsymbol{g}(\boldsymbol{x}, \boldsymbol{u}, \boldsymbol{u}') = \boldsymbol{g}, \text{ where } \boldsymbol{u} = V_{ss},\ \boldsymbol{u}' = [V\ f]^T \text{ or } \boldsymbol{u}' = [V\ \theta]^T, \tag{2.1}$$

$$\boldsymbol{x} = [\alpha\ \omega\ E_d'\ E_q'\ \Psi_{1d}\ \Psi_{2q}\ E_{dc}'\ V_r]^T;\ \text{or},\ \boldsymbol{x} = [\delta\ \omega\ E_d'\ E_q'\ \Psi_{1d}\ \Psi_{2q}\ E_{dc}'\ V_r]^T. \tag{2.2}$$

In the definition of state vector $\boldsymbol{x}$ (in (2.2)) either α or δ can be taken as a state. The vector $\boldsymbol{g}$ consists of eight functions corresponding to the eight states in $\boldsymbol{x}$, $\boldsymbol{g} = [g_\alpha\ g_\omega\ g_{E_d'}\ g_{E_q'}\ g_{\Psi_{1d}}\ g_{\Psi_{2q}}\ g_{E_{dc}'}\ g_{V_r}]^T$ or $\boldsymbol{g} = [g_\delta\ g_\omega\ g_{E_d'}\ g_{E_q'}\ g_{\Psi_{1d}}\ g_{\Psi_{2q}}\ g_{E_{dc}'}\ g_{V_r}]^T$ (assuming the case of an ST1A excitation system). The DAEs for a generating unit are as follows:

$$\dot{\alpha} = \omega_b(\omega - f) = g_\alpha, \tag{2.3}$$

$$\dot{\delta} = \omega_b(\omega - 1) = g_\delta, \tag{2.4}$$

$$\dot{\omega} = (T_m - T_e - D(\omega - 1))/(2H) = g_\omega, \tag{2.5}$$

$$\dot{E}'_d = -(E'_d + (X_q - X'_q)\{K_{q1}I_q + K_{q2}\frac{\Psi_{2q} + E'_d}{X'_q - X_l}\})/T'_{q0} = g_{E'_d}, \tag{2.6}$$

$$\dot{E}'_q = [E_{fd} + (X_d - X'_d)\{K_{d1}I_d + K_{d2}\frac{\Psi_{1d} - E'_q}{X'_d - X_l}\} - E'_q]/T'_{d0} = g_{E'_q}, \tag{2.7}$$

$$\dot{\Psi}_{1d} = (E'_q + (X'_d - X_l)I_d - \Psi_{1d})/T''_{d0} = g_{\Psi_{1d}}, \tag{2.8}$$

$$\dot{\Psi}_{2q} = (-E'_d + (X'_q - X_l)I_q - \Psi_{2q})/T''_{q0} = g_{\Psi_{2q}}, \tag{2.9}$$

$$\dot{E}'_{dc} = ((X''_d - X''_q)I_q - E'_{dc})/T_c = g_{E'_{dc}}, \tag{2.10}$$

$$I_d = (R_a(E'_d K_{q1} - \Psi_{2q}K_{q2} + E'_{dc} - V_d) - X''_d(E'_q K_{d1} + \Psi_{1d}K_{d2} - V_q))/Z_a^2, \tag{2.11}$$

$$I_q = (R_a(E'_q K_{d1} + \Psi_{1d}K_{d2} - V_q) + X''_d(E'_d K_{q1} - \Psi_{2q}K_{q2} + E'_{dc} - V_d))/Z_a^2, \tag{2.12}$$

$$\text{also, } I_q = I\cos(\delta - \phi),\ I_d = -I\sin(\delta - \phi),\ I\angle{-\phi} = (P + jQ)/(V\angle\theta), \tag{2.13}$$

$$V_q = V\cos\alpha = V\cos(\delta - \theta),\ V_d = -V\sin\alpha = -V\sin(\delta - \theta), \tag{2.14}$$

$$T_e = E'_d I_d K_{q1} + E'_q I_q K_{d1} + \Psi_{1d} I_q K_{d2} - \Psi_{2q} I_d K_{q2} + I_d I_q (X''_d - X''_q). \tag{2.15}$$

The equations defining an IEEE-ST1A excitation system are given as follows:

$$\dot{V}_r = (V - V_r)/T_r = g_{V_r}, \tag{2.16}$$

$$\text{where } E_{fd} = K_a(V_{ref} + V_{ss} - V_r),\ E_{fdmin} \le E_{fd} \le E_{fdmax}. \tag{2.17}$$

The equations defining an IEEE-DC1A excitation system, in which feedback of excitation voltage is not present, are given as follows:

$$\dot{V}_r = (V - V_r)/T_r = g_{V_r}, \tag{2.18}$$

$$\dot{V}_a = [K_a(V_{ref} + V_{ss} - V_r) - V_a]/T_a = g_{V_a}, \tag{2.19}$$

$$\dot{E}_{fd} = -[E_{fd}(K_x + A_x e^{B_x E_{fd}}) - V_a]/T_x = g_{E_{fd}}, \text{ where } V_{amin} \le V_a \le V_{amax}. \quad (2.20)$$

Note that the vectors $\boldsymbol{x}$ and $\boldsymbol{g}$ will change if a DC1A excitation system is used in the generating unit, instead of an ST1A excitation system.

2.1.2 Power system stabilizers (PSSs)

PSSs are used to provide supplementary signals to stabilize power system dynamics in the postdisturbance phase. Usually, a high-gain fast acting type of excitation control system provides good transient stability. The fast excitation control at times also contributes to the oscillatory response of the system.

Sometimes the system is (or becomes) ill-conditioned with poorly damped electromechanical modes (the modes of the system are described in the next section). Postdisturbance oscillations for such a system may grow out of control, and this requires additional stabilization. PSSs provide such a stabilization and are used as supplements to damp the electromechanical modes of the system. The feedback signal to a PSS may be the rotor speed (or slip), terminal voltage of the generator, real power, or reactive power generated; such a signal is chosen which has strong observability of the modes. Typically, when the rotor slip $(\omega - 1)$ is used as the feedback signal, the dynamic equation of a PSS with two lead–lag stages is given as follows:

$$V_{ss} = K_{pss} \frac{(sT_w)}{(1 + sT_w)} \frac{(1 + sT_{11})}{(1 + sT_{12})} \frac{(1 + sT_{21})}{(1 + sT_{22})} (\omega - 1). \quad (2.21)$$

2.1.3 FACTS control devices

As mentioned in Chapter 1, FACTS control devices are installed in a power system to exercise continuous control over the bus voltages or line power flows [19]. Within the economic generation dispatch schedule, these devices enable the voltages and power flows to be controlled without exceeding the operating limits and at the same time increasing stability margins and minimizing losses. Some of the important FACTS control devices are described as follows.

2.1.3.1 Thyristor-controlled series capacitor (TCSC)

A TCSC is a controllable FACTS control device which is used for dynamic power flow control through variation of the reactance of a transmission tie line connecting two areas, in order to dampen the electromechanical modes

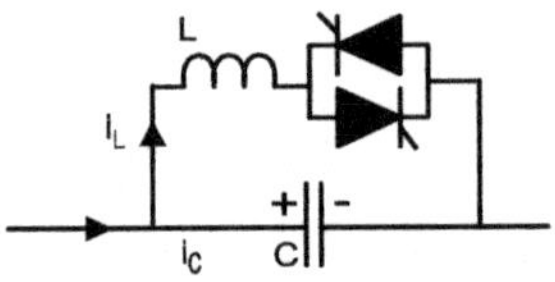

Figure 2.3 Topology of a TCSC.

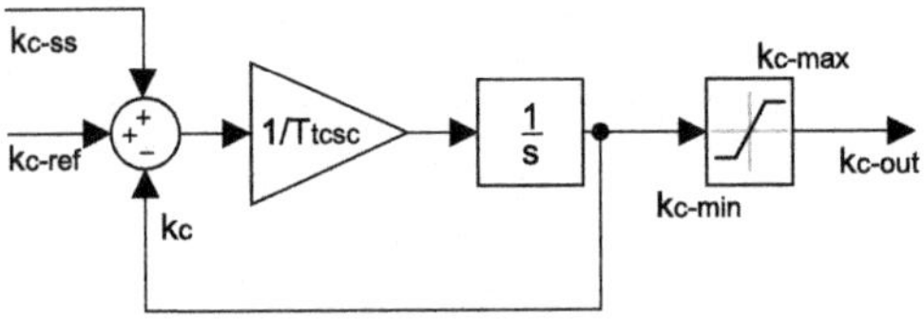

Figure 2.4 Block diagram of TCSC dynamic model.

by facilitating power transfer across the tie line. The topology of a TCSC is shown in Fig. 2.3.

The control action of a TCSC is expressed in terms of its percentage compensation $K_c = \frac{X_C}{X_L} \times 100\%$, where X_L is the reactance of the line and X_C is the capacitive reactance offered by the TCSC, thereby making the effective reactance of the line as $(X_L - X_C)$. The dynamic characteristic of the TCSC are assumed to be modeled by a single and fast time constant T_{tcsc}, and its dynamics are given by the following equation, with block diagram as in Fig. 2.4:

$$T_{tcsc}\dot{K}_c = -K_c + K_{c-ref} + K_{c-ss}. \tag{2.22}$$

2.1.3.2 Static VAR compensator (SVC)

An SVC is a shunt-connected static device which exchanges capacitive or inductive current in order to control the voltage (or any other variable) of the bus to which it is connected. It consists of a thyristor-controlled reactor in parallel to a fixed capacitor. An SVC functions by varying the susceptance across the bus or, equivalently, through reactive power injection at the bus.

The reactive power injection of an SVC, denoted by Q_{svc}, and the dynamic model of an SVC are given as follows. The terms B_C and B_L are the susceptance of the capacitor and the thyristor-controlled reactor, respectively, T_{svc} is the switching response time, T_{mes} is measurement delay, T_{v1} and T_{v2} are time constants of the voltage regulator of the SVC, K_v is the regulator gain of the SVC, V_{r-svc} and V_{t-svc} are dynamic voltages corresponding to the regulator and measurement stages of the SVC, respectively,

$V_{ref-svc}$ is the reference voltage input given to the SVC, and V_{ss-svc} is the small-signal control input given to the SVC. We have

$$Q_{svc} = V^2 B_{svc}, \quad B_{svc} = B_C - B_L, \tag{2.23}$$

$$T_{svc}\dot{B}_{svc} = -B_{svc} + \left(1 - \frac{T_{v1}}{T_{v2}}\right) V_{r-svc} + \frac{K_v T_{v1}}{T_{v2}}[V_{ss-svc} + V_{ref-svc} - V_{t-svc}], \tag{2.24}$$

$$T_{v2}\dot{V}_{r-svc} = -V_{r-svc} - K_v V_{t-svc} + K_v[V_{ss-svc} + V_{ref-svc}], \tag{2.25}$$

$$T_{mes}\dot{V}_{t-svc} = -V_{t-svc} + V. \tag{2.26}$$

2.1.3.3 Thyristor-controlled phase angle regulator (TCPAR)

A thyristor-controlled phase angle regulator (TCPAR) is a phase shifting transformer controlled by thyristor switches to dynamically vary the phase angle of one of the phases of a three-phase line to which the TCPAR is connected in series. The phase shifting is done by adding a voltage vector in series with the line impedance of the phase. This voltage vector is derived from the other two phases via a shunt-connected transformer.

The voltage vector is assumed to be modeled by an ideal voltage source, $\bar{V}_{sc} = V_s \angle \phi_{tcpar}$, and the dynamic equation governing ϕ_{tcpar} is given as follows. The term T_{tcpar} is the switching response time of the TCPAR, $\phi_{tcpar-ref}$ is the reference input given to the TCPAR, and $\phi_{tcpar-ss}$ is the small-signal control input given to the TCPAR. We have

$$T_{tcpar}\dot{\phi}_{tcpar} = -\phi_{tcpar} + \phi_{tcpar-ref} + \phi_{tcpar-ss}. \tag{2.27}$$

2.1.4 Loads, network interface, and network equations

While writing the algebraic equations of the network, also called network balance equations, all the equations should be in a common reference frame of the network, instead of the rotating reference frame of the generators. Therefore the generator currents and voltages need to be rotated by the rotor phase angle δ. The resulting equations are given as follows:

$$I_{Qi} + jI_{Di} = (I_{qi} + jI_{di})e^{j\delta_i}; \text{ and } V_{gi} = V_{Qi} + jV_{Di} = (V_{qi} + jV_{di})e^{j\delta_i}. \tag{2.28}$$

Here, the variables with upper-case subscripts (that is, Q and D) are network variables, while those with lower-case subscripts (that is, q and d) are machine variables. In order to club the generator admittances with the network admittance, the generator is represented as a current injection source,

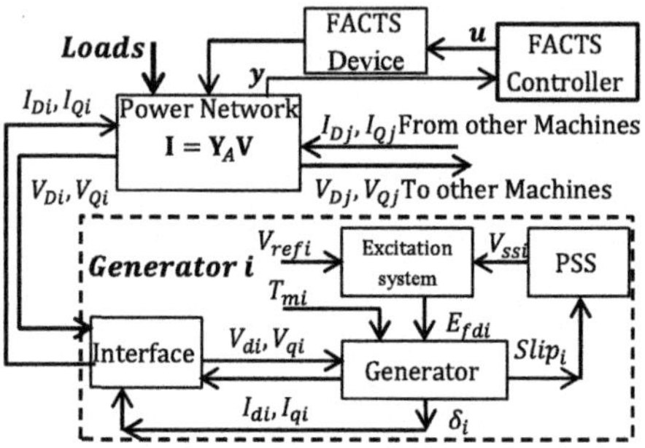

Figure 2.5 Block diagram of a typical power system.

given by the following equation:

$$I_{gi} = (I_{Qi} + jI_{Di}) + Y_{gi} V_{gi}, \tag{2.29}$$

where $Y_{gi} = \frac{1}{(R_{ai}+jX''_{di})}$. The generator admittance matrix, $\mathbf{Y}_G$, is equal to diag$\{[Y_{G1}, Y_{G2}, \ldots, Y_{GN}]\}$, where N is the total number of bus nodes and $Y_{Gj} = Y_{gi}$ if the jth node is connected to the ith generator, and $Y_{Gj} = 0$ otherwise. Similarly, the load admittance matrix $\mathbf{Y}_L$ is defined, where loads are taken as constant shunt impedances. Assuming that the current injections take place only at the generator buses, the current injection column vector $\mathbf{I}$ is such that its jth element $I_j = I_{gi}$ if the jth node is connected to the ith generator, and $I_j = 0$ otherwise, for $j = 1$ to N. The bus admittance matrix $\mathbf{Y}_N$ is formed using the line impedances as follows:

$$\mathbf{Y}_{N,ij} = \begin{cases} y_i + \sum_{k=1,2,\ldots,N \ \& \ k \neq i} y_{ik}, & \text{if} \quad i = j, \\ -y_{ij}, & \text{if} \quad i \neq j. \end{cases} \tag{2.30}$$

Here, $\mathbf{Y}_{N,ij}$ denotes the (i,j)th element of $\mathbf{Y}_N$, y_i denotes the admittance of the ith bus (excluding the load and generator admittances), and y_{ij} denotes the admittance between bus i and bus j. Once $\mathbf{Y}_N$ is calculated, it is augmented with $\mathbf{Y}_G$ and $\mathbf{Y}_L$ to give $\mathbf{Y}_A = \mathbf{Y}_N + \mathbf{Y}_G + \mathbf{Y}_L$, and the column vector of bus voltages $\mathbf{V}$ is given as follows:

$$\mathbf{V} = \mathbf{Z}_A I, \text{ where } \mathbf{Z}_A = (\mathbf{Y}_A)^{-1}. \tag{2.31}$$

The integrated block diagram of an interconnected power system with different components as discussed in this section is shown in Fig. 2.5.

2.2 POWER SYSTEM SIMULATION AND ANALYSIS

2.2.1 Load flow analysis

The first step in power system simulation is to find the initial steady-state voltage magnitudes and phases for all the buses (or the nodes) in the system, and this step is called load flow or power flow solution. For finding the initial bus voltage magnitudes and phases, (V_i, θ_i), $1 \leq i \leq N$, the following load flow equations need to be solved, which are an alternate form of (2.31):

$$P_i = \sum_{j=1}^{N} V_i V_j [G_{ij} \cos(\theta_i - \theta_j) + B_{ij} \sin(\theta_i - \theta_j)],$$
$$Q_i = \sum_{j=1}^{N} V_i V_j [G_{ij} \sin(\theta_i - \theta_j) - B_{ij} \cos(\theta_i - \theta_j)]. \tag{2.32}$$

Here, $\boldsymbol{G}$ and $\boldsymbol{B}$ respectively denote the real and imaginary parts of the aforementioned bus admittance matrix, $\boldsymbol{Y}_N$, while P_i and Q_i respectively denote the net real and reactive powers injected at bus i. The data needed for solving these load flow equations are the following.

1. *The bus admittance matrix.* This matrix is computed using (2.30), which requires data of line impedances (for the test system presented in Appendix A, the line data are given in Table A.2). Note that the steady-state value of the bus admittance matrix also depends on the initial setting of the FACTS control device in the system.
2. *The bus voltage magnitude and phase for bus corresponding to one of the generators.* This bus is called the slack bus or the swing bus (as an example, the bus numbered 16 for the 16-machine, 68-bus test system described in Appendix A is the swing bus, with its "bus type" specified as 1 in Table A.1).
3. *The bus voltage magnitudes and injected real power for the rest of the generator buses.* These buses are called PV buses (for example, buses 1 to 15 of the test system in Appendix A are PV buses, with "bus type" as specified as 2 in Table A.1).
4. *The real and reactive power consumption of nongenerator buses.* These buses are called PQ buses (for example, buses 17 to 68 of the test system in Appendix A are PQ buses, with "bus type" specified as 3 in Table A.1).

The reactive powers for generator buses are not known and can only be computed after finding all the bus voltage magnitudes and phases. Hence, for the reactive part of (2.32), that is, for $Q_i = \sum_{j=1}^{N} V_i V_j [G_{ij} \sin(\theta_i - \theta_j) - B_{ij} \cos(\theta_i - \theta_j)]$, only $(N - M)$ equations corresponding to the PQ buses (with $(M + 1) \leq i \leq N$) are used in finding the bus voltage magnitudes and phases, whereas the rest of the M equations (with $1 \leq i \leq M$) are used after finding the bus voltage magnitudes and phases, in order to calculate Q for generator buses. Similarly, the swing bus real power is not known and, hence, $(N - 1)$ real power equations are utilized in finding the bus voltage magnitudes and phases, while the remaining one equation is utilized in finding the swing bus real power, once all the bus voltage magnitudes and phases have been computed.

Thus, there are a total of $(2N - M - 1)$ unknowns (that is, $(N - M)$ unknown voltage magnitudes for PQ buses and $(N - 1)$ unknown voltage phases for nonswing buses) and $(2N - M - 1)$ equations (that is, $(N - 1)$ real power equations and $(N - M)$ reactive power equations). As the number of equations is equal to the number of unknown variables, the load flow equations can be solved if a solution exists. However, solving these equations is not straightforward as the equations are nonlinear.

Different methods have been employed for solving the load flow equations. The most common of these methods is the Newton–Raphson method, which is an iterative method based on Taylor series and initial guesses for bus voltage magnitudes and phases. We start with the initial guesses; either they are specified in the bus data, or a "flat" guess is used, with zero as the initial guess for each of the voltage phases (except for the swing bus, for which the phase is always specified in the data) and 1 p.u. as the initial guess for the voltage magnitude of each of the nongenerator buses (or the PQ buses). The next step is to find the first-order Taylor expansion of the $(2N - M - 1)$ load flow equations, which is given as follows:

$$\begin{bmatrix} \Delta\boldsymbol{\theta} \\ \Delta\boldsymbol{V} \end{bmatrix} = \boldsymbol{J}^{-1} \begin{bmatrix} \Delta\boldsymbol{P} \\ \Delta\boldsymbol{Q} \end{bmatrix}, \quad \boldsymbol{J} = \begin{bmatrix} \frac{\partial \Delta\boldsymbol{P}}{\partial \boldsymbol{\theta}} & \frac{\partial \Delta\boldsymbol{P}}{\partial \boldsymbol{V}} \\ \frac{\partial \Delta\boldsymbol{Q}}{\partial \boldsymbol{\theta}} & \frac{\partial \Delta\boldsymbol{Q}}{\partial \boldsymbol{V}} \end{bmatrix}. \tag{2.33}$$

Here, $\Delta\boldsymbol{P}$ is the column vector of ΔP_i for nonswing buses and $\Delta\boldsymbol{Q}$ is the column vector of ΔQ_i for PQ buses, defined as follows:

$$\begin{aligned} \Delta P_i &= -P_i + \sum_{j=1}^{N} V_i V_j [G_{ij} \cos(\theta_i - \theta_j) + B_{ij} \sin(\theta_i - \theta_j)], \\ \Delta Q_i &= -Q_i + \sum_{j=1}^{N} V_i V_j [G_{ij} \sin(\theta_i - \theta_j) - B_{ij} \cos(\theta_i - \theta_j)], \end{aligned} \tag{2.34}$$

where P_i and Q_i are the total real and reactive power injections at the ith bus. Also, $\Delta\boldsymbol{\theta}$ is the column vector of $\Delta\theta_i$ for nonswing buses, $\Delta\boldsymbol{V}$ is the column vector of ΔV_i for PQ buses, and their $(k+1)$th iteration values are used to correct the kth iteration guesses for V_i and θ_i as follows:

$$V_i^{k+1} = \Delta V_i^{k+1} + V_i^k, \quad \theta_i^{k+1} = \Delta\theta_i^{k+1} + \theta_i^k. \tag{2.35}$$

The iterations of Newton–Raphson method are stopped once the magnitudes of $\Delta\boldsymbol{\theta}$ and $\Delta\boldsymbol{V}$ become less than a small threshold value which is close to zero. When the iterations continue beyond a maximum specified number (typically 10–12), the method fails to converge. This can either mean that a solution does not exist, or that it cannot be found using the Newton–Raphson method. In these rare cases, other methods for solving the load flow equations, such as the Gauss–Seidel method, fast decoupled load flow method, or holomorphic embedding load flow method, can be used. Since in most cases when a solution exists the Newton–Raphson method converges, a discussion on the alternate methods is not pursued in this text.

Load flow analysis has been performed on the 16-machine, 68-bus test system described in Appendix A. The bus data used are given in Table A.1, while the bus admittance matrix is calculated through (2.30) using the line data of Table A.2, after including a TCSC on the line between buses 18 and 50, using parameters specified in A.1.6. A "flat" initial guess is used: 1 p.u. for voltage magnitudes of the nongenerator buses (that is, buses 17 to 68) and 0 rad for voltage phases of nonswing buses (that is, all buses except the 16th bus). The method converges in five iterations and Table 2.1 presents the results. The code – without TCSC – for the Newton–Raphson method is available online at [134] ([134] also contains other simulation codes for the test system of Appendix A, with a slightly modified excitation system and PSS parameters).

2.2.2 Initialization and time-domain simulation

Initial steady-state conditions of the system are found by first finding the initial bus voltage magnitudes and phases using load flow calculation (as described in Subsection 2.2.1), and then using these values to find the steady-state values of the system states using the DAEs described in Section 2.1. This is done by making all the derivative terms in the DAEs equal to zero (as the derivative terms are zero in steady state) and then finding the value of the states using the resultant algebraic equations. Note that

Table 2.1 Load flow for the 68-bus system

Bus no.	V (p.u.)	θ (degree)	P_G (p.u.)	Q_G (p.u.)	P_L (p.u.)	Q_L (p.u.)	Bus type
1	1.045	−5.095	2.5	1.947	0	0	2
2	0.98	2.956	5.45	0.691	0	0	2
3	0.983	5.543	6.5	0.801	0	0	2
4	0.997	5.565	6.32	0.001	0	0	2
5	1.011	3.269	5.05	1.165	0	0	2
6	1.05	7.739	7	2.542	0	0	2
7	1.063	9.927	5.6	2.907	0	0	2
8	1.03	1.025	5.4	0.483	0	0	2
9	1.025	6.528	8	0.595	0	0	2
10	1.01	−5.896	5	−0.169	0	0	2
11	1	−3.161	10	0.072	0	0	2
12	1.016	−18.389	13.5	2.831	0	0	2
13	1.011	−24.378	35.91	8.934	0	0	2
14	1	12.676	17.85	0.351	0	0	2
15	1	0.873	10	0.785	0	0	2
16	1	0	33.711	−0.02	0	0	1
17	0.95	−31.753	0	0	60	3	3
18	1.005	−5.775	0	0	24.7	1.23	3
19	0.932	−0.367	0	0	0	0	3
20	0.981	−1.978	0	0	6.8	1.03	3
21	0.96	−3.157	0	0	2.74	1.15	3
22	0.994	2.095	0	0	0	0	3
23	0.996	1.736	0	0	2.48	0.85	3
24	0.959	−5.98	0	0	3.09	−0.92	3
25	0.998	−6.132	0	0	2.24	0.47	3
26	0.987	−7.143	0	0	1.39	0.17	3
27	0.968	−8.976	0	0	2.81	0.76	3
28	0.99	−3.621	0	0	2.06	0.28	3
29	0.992	−0.671	0	0	2.84	0.27	3
30	0.977	−15.811	0	0	0	0	3
31	0.985	−13.708	0	0	0	0	3
32	0.97	−11.173	0	0	0	0	3
33	0.974	−15.635	0	0	1.12	0	3
34	0.979	−21.731	0	0	0	0	3
35	1.039	−22.534	0	0	0	0	3
36	0.96	−24.586	0	0	1.02	−0.195	3
37	0.956	−7.897	0	0	0	0	3

continued on next page

Table 2.1 *(continued)*

Bus no.	V (p.u.)	θ (degree)	P_G (p.u.)	Q_G (p.u.)	P_L (p.u.)	Q_L (p.u.)	Bus type
38	0.99	−15.14	0	0	0	0	3
39	0.986	−34.701	0	0	2.67	0.126	3
40	1.05	−10.609	0	0	0.656	0.235	3
41	1	11.141	0	0	10	2.5	3
42	0.999	0.013	0	0	11.5	2.5	3
43	0.973	−33.425	0	0	0	0	3
44	0.974	−33.494	0	0	2.675	0.048	3
45	1.038	−24.418	0	0	2.08	0.21	3
46	0.993	−17.488	0	0	1.507	0.285	3
47	1.022	−15.957	0	0	2.031	0.326	3
48	1.038	−15.019	0	0	2.412	0.022	3
49	0.999	−17.337	0	0	1.64	0.29	3
50	1.044	−12.845	0	0	1	−1.47	3
51	1.051	−21.984	0	0	3.37	−1.22	3
52	0.955	−8.944	0	0	1.58	0.3	3
53	0.988	−15.145	0	0	2.527	1.186	3
54	0.986	−7.675	0	0	0	0	3
55	0.957	−9.33	0	0	3.22	0.02	3
56	0.921	−8.038	0	0	2	0.736	3
57	0.91	−7.271	0	0	0	0	3
58	0.909	−6.46	0	0	0	0	3
59	0.904	−9.357	0	0	2.34	0.84	3
60	0.907	−10.077	0	0	2.088	0.708	3
61	0.956	−19.168	0	0	1.04	1.25	3
62	0.912	−3.38	0	0	0	0	3
63	0.91	−4.435	0	0	0	0	3
64	0.837	−4.445	0	0	0.09	0.88	3
65	0.913	−4.256	0	0	0	0	3
66	0.92	−6.27	0	0	0	0	3
67	0.928	−7.525	0	0	3.2	1.53	3
68	0.948	−6.174	0	0	3.29	0.32	3

the initial rotor angle, δ_0, and initial excitation voltage, E_{fd0}, cannot be found using this method; instead, they are found by defining a new quantity $E_q\angle\delta = \left(V\angle\theta + (R_a + jX_q)I\angle\phi\right)$ [63].

Thus, the initial states for a generating unit modeled using DAEs in (2.3)–(2.17) are given as follows (with V_0, θ_0, P_0, and Q_0 for the generation unit obtained from load flow analysis):

$$I_0\angle{-\phi_0} = (P_0 + jQ_0)/(V_0\angle\theta_0), \tag{2.36}$$

$$E_{q0}\angle\delta_0 = \left(V_0\angle\theta_0 + (R_a + jX_q)I_0\angle\phi_0\right), \tag{2.37}$$

$$\omega_0 = 1, \tag{2.38}$$

$$I_{q0} = I_0\cos(\delta_0 - \phi_0),\ I_{d0} = -I_0\sin(\delta_0 - \phi_0), \tag{2.39}$$

$$V_{q0} = V_0\cos\alpha_0 = V_0\cos(\delta_0 - \theta_0),\ V_{d0} = -V_0\sin\alpha_0 = -V_0\sin(\delta_0 - \theta_0), \tag{2.40}$$

$$E_{fd0} = E_{q0} - (X_d - X_q)I_{d0}, \tag{2.41}$$

$$E'_{q0} = E_{fd0} + (X_d - X'_d)I_{d0}, \tag{2.42}$$

$$E'_{d0} = -(X_q - X'_q)I_{q0}, \tag{2.43}$$

$$\Psi_{1d0} = E'_{q0} + (X'_d - X_l)I_{d0}, \tag{2.44}$$

$$\Psi_{2q0} = -E'_{d0} + (X'_q - X_l)I_{q0}, \tag{2.45}$$

$$E'_{dc0} = (X''_d - X''_q)I_{q0}, \tag{2.46}$$

$$V_{r0} = V_0,\quad V_{ref} = (E_{fd0} + V_0)/K_a, \tag{2.47}$$

$$T_{m0} = T_{e0} = P_0. \tag{2.48}$$

The initial states for other system states (such as for PSSs and FACTS control devices) are similarly obtained. Once the initial states are obtained, the DAEs given in Section 2.1 can be implemented in simulation software (such as MATLAB-Simulink) to perform time–domain simulations of a power system model. Fig. 2.6 shows an example plot for a time–domain simulation of the 68-bus test system of Appendix A (readers interested in coding may refer to [134] for a guide to the MATLAB-Simulink codes).

In the simulated system, generating units 1 to 8 have a DC1A excitation system, while unit 9 has an ST1A static excitation system and also has a PSS. Also, a TCSC is present on the line between buses 18 and 50. In the simulation, the system starts from a steady state and at 1 second a disturbance is introduced in the system by disconnecting one of the tie lines between buses 53 and 54. From Fig. 2.6 it can be observed that the system appears to be at the margin of stability and instability, as the oscillations which ensue after the disturbance damp out, but very slowly. More accurate

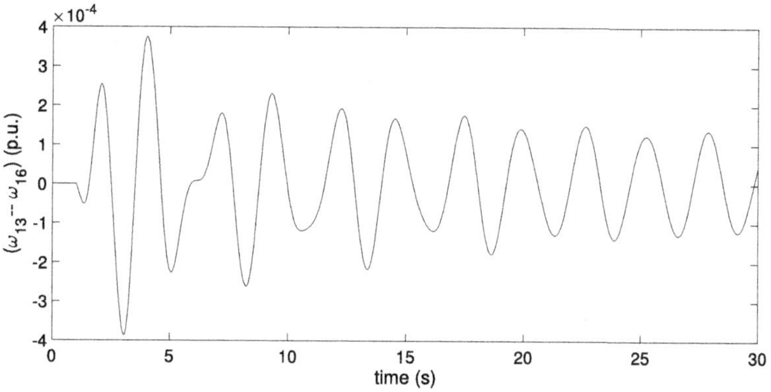

Figure 2.6 Plot of ω_{13}–ω_{16} for the 68-bus test system.

information about the stability of the system can be obtained by performing its linear analysis, as described in the next subsection.

2.2.3 Linear analysis and basics of control design

2.2.3.1 System linearization

Let the (nonlinear) DAEs of a power system be represented as follows:

$$\begin{aligned} \dot{x} &= \bar{g}(x, u, y_a), \\ 0 &= \bar{h}(x, u, y_a), \\ y &= h(x, u, y_a). \end{aligned} \tag{2.49}$$

For linear analysis (or small-signal analysis), the system of DAEs given by (2.49) are first linearized at a steady-state operating point using the calculated steady-state values, and then state-space matrices and eigenvalues for the linearized system are obtained. If the steady-state values of x, u, y_a, and y are denoted by x_0, u_0, y_{a0}, and y_0, respectively, with $\Delta x = x - x_0$, $\Delta u = u - u_0$, $\Delta y_a = y_a - y_{a0}$, and $\Delta y = y - y_0$, the linearization of (2.49) about the steady-state values is given as follows:

$$\begin{aligned} \Delta \dot{x} &= \frac{\partial \bar{g}}{\partial x}\Delta x + \frac{\partial \bar{g}}{\partial u}\Delta u + \frac{\partial \bar{g}}{\partial y_a}\Delta y_a, \\ 0 &= \frac{\partial \bar{h}}{\partial x}\Delta x + \frac{\partial \bar{h}}{\partial u}\Delta u + \frac{\partial \bar{h}}{\partial y_a}\Delta y_a, \\ \Delta y &= \frac{\partial h}{\partial x}\Delta x + \frac{\partial h}{\partial u}\Delta u + \frac{\partial h}{\partial y_a}\Delta y_a. \end{aligned} \tag{2.50}$$

All the partial derivatives in (2.50) are obtained at $\boldsymbol{x} = \boldsymbol{x}_0$, $\boldsymbol{u} = \boldsymbol{u}_0$, $\boldsymbol{y}_a = \boldsymbol{y}_{a0}$, and $\boldsymbol{y} = \boldsymbol{y}_0$. If the dimensions of $\boldsymbol{y}_a$ and $\bar{\boldsymbol{h}}$ are equal (such that the number of algebraic equations is equal to the number of algebraic variables), $\boldsymbol{y}_a$ can be eliminated from (2.50), giving the following state–space form:

$$\begin{aligned} \Delta\dot{\boldsymbol{x}} &= \boldsymbol{A}\Delta\boldsymbol{x} + \boldsymbol{B}\Delta\boldsymbol{u}, \\ \Delta\boldsymbol{y} &= \boldsymbol{C}\Delta\boldsymbol{x} + \boldsymbol{D}\Delta\boldsymbol{u}, \end{aligned} \tag{2.51}$$

$$\begin{aligned} \boldsymbol{A} &= \frac{\partial\bar{\boldsymbol{g}}}{\partial\boldsymbol{x}} - \frac{\partial\bar{\boldsymbol{g}}}{\partial\boldsymbol{y}_a}\left(\frac{\partial\bar{\boldsymbol{h}}}{\partial\boldsymbol{y}_a}\right)^{-1}\frac{\partial\bar{\boldsymbol{h}}}{\partial\boldsymbol{x}}, \quad \boldsymbol{B} = \frac{\partial\bar{\boldsymbol{g}}}{\partial\boldsymbol{u}} - \frac{\partial\bar{\boldsymbol{g}}}{\partial\boldsymbol{y}_a}\left(\frac{\partial\bar{\boldsymbol{h}}}{\partial\boldsymbol{y}_a}\right)^{-1}\frac{\partial\bar{\boldsymbol{h}}}{\partial\boldsymbol{u}}, \\ \boldsymbol{C} &= \frac{\partial\boldsymbol{h}}{\partial\boldsymbol{x}} - \frac{\partial\boldsymbol{h}}{\partial\boldsymbol{y}_a}\left(\frac{\partial\bar{\boldsymbol{h}}}{\partial\boldsymbol{y}_a}\right)^{-1}\frac{\partial\bar{\boldsymbol{h}}}{\partial\boldsymbol{x}}, \quad \boldsymbol{D} = \frac{\partial\boldsymbol{h}}{\partial\boldsymbol{u}} - \frac{\partial\boldsymbol{h}}{\partial\boldsymbol{y}_a}\left(\frac{\partial\bar{\boldsymbol{h}}}{\partial\boldsymbol{y}_a}\right)^{-1}\frac{\partial\bar{\boldsymbol{h}}}{\partial\boldsymbol{u}}. \end{aligned} \tag{2.52}$$

After linearizing the system about the steady-state values, the next step is to find the eigenvalues of the linear system. Simulink has inbuilt functions for both linearizing a system to find its state–space matrices and finding eigenvalues using the system's state–transition matrix, $\boldsymbol{A}$.

2.2.3.2 Eigenvalues

An eigenvalue of a dynamic system which can be represented in form of (2.51) is defined as a root of the equation $(\boldsymbol{A} - \lambda\boldsymbol{I}) = \boldsymbol{0}$. If λ_i is the ith eigenvalue of the system, then the right eigenvector, $\boldsymbol{r}_i$, and the left eigenvector, $\boldsymbol{l}_i$, corresponding to λ_i are given by the equations $(\boldsymbol{A}\boldsymbol{r}_i - \lambda_i\boldsymbol{r}_i) = \boldsymbol{0}$ and $(\boldsymbol{l}_i\boldsymbol{A} - \lambda_i\boldsymbol{l}_i) = \boldsymbol{0}$, respectively. The frequency of an eigenvalue (in rad/s) is given by the absolute value of its imaginary part (in Hz, it is given by [absolute value of imaginary part]$/2\pi$), while the damping of an eigenvalue is given by the negative of its real part. The damping ratio, ζ, of an eigenvalue is defined as the ratio of the damping of the eigenvalue to its absolute value.

Eigenvalues, also called modes, are directly related to the stability of a linear system. Any dynamic system (including a power system) is said to be small-signal stable about an initial steady-state operating point if the system reaches a steady state after it is subjected to a small disturbance. This happens if and only if all the real parts of the eigenvalues of the linearized system are negative, that is, all the eigenvalues lie to the left of the imaginary axis in the complex plane.

Another condition which must be satisfied by the eigenvalues of a power system is that the damping ratio of low-frequency eigenvalues of the system

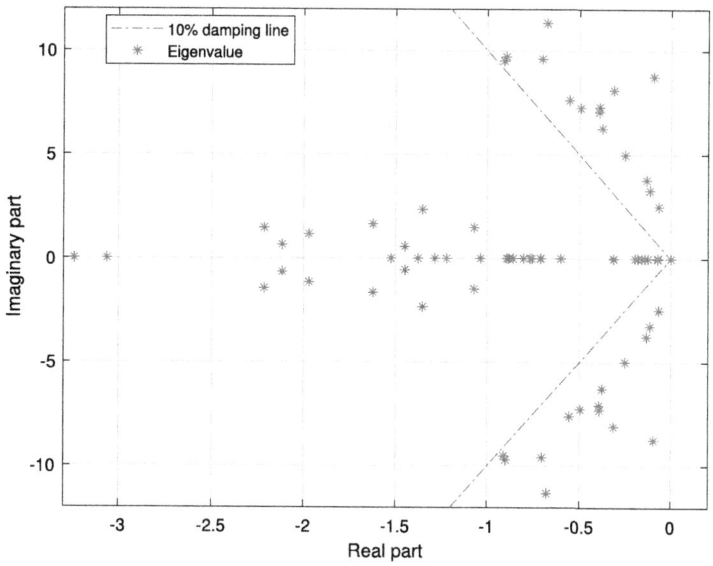

Figure 2.7 Plot of the eigenvalues of the 68-bus test system.

should be more than a certain percentage (usually 10% to 15%), so that any system oscillation damps out quickly, without causing any stress to the devices present in the system. This means that the ratio of the negative of the real part of the eigenvalue to its absolute value should be more than 0.1, if the minimum required damping ratio is, say, 10%.

Fig. 2.7 shows the plot of the eigenvalues of the 68-bus test system. The system is small-signal stable as all the eigenvalues have negative real parts, but it is poorly damped due to many of its eigenvalues being outside the 10% damping conic section, especially those which have an absolute of the imaginary part less than 2π (that is, with frequencies less than 1 Hz).

2.2.3.3 Participation factor and residue

Control design, or the design of controllers, refers to techniques for controlling the modes of a system using any controllable device in the system, such as a generating unit or a FACTS device. A detailed modal analysis of the modes of a power system is required to design linear control methods. Modal analysis includes participation factor analysis and residue analysis. The participation factor shows the extent of participation of different states of the system in a given mode. The participation factor of the jth state in the ith mode is defined as follows:

$$PF_{ij} = \boldsymbol{r}_{ij}\boldsymbol{l}_{ij}, \tag{2.53}$$

where $\boldsymbol{r}_i$ and $\boldsymbol{l}_i$ denote the right and left eigenvectors of the ith mode, respectively. The generating unit (or any other controllable device in the system, such as a TCSC) which has states with the highest participation factors in a given mode of interest, is a good candidate for control design for controlling that mode.

Although the participation factor gives a good idea of which devices should be used, it does not tell anything about which input/output pair of signals should be used for the control design. A factor which helps in deciding which signals should act as the input/output pair for control design is the "residue", which indicates the product of observability and controllability of any given pair of input and output signals. The residue associated with the jth pair of input/output and the ith mode is defined as follows:

$$R_{ij} = \boldsymbol{C}_j\boldsymbol{r}_i\boldsymbol{l}_i\boldsymbol{B}_j, \tag{2.54}$$

where $\boldsymbol{C}_j$ is that *row* of $\boldsymbol{C}$ which corresponds to the output in the jth input/output pair for the system, while $\boldsymbol{B}_j$ is that *column* of $\boldsymbol{B}$ which corresponds to the input in the jth input/output pair for the system. Note that the absolute value of the residue of a given pair of input and output is equal to the product of the pair's observability and controllability. The input/output pairs with high residues in a given mode are good candidates for controlling that mode.

2.2.3.4 Controllable devices, controllers, inputs, and outputs: examples

Two examples of controllable devices in a power system are the excitation system of a machine and a TCSC (or any other FACTS control device). Corresponding examples of controllers are a PSS, for controlling the excitation produced by the excitation system, and a power oscillation damper (POD), for controlling the compensation of a TCSC. Examples of inputs to the controllable devices are the V_{ss} signal (which is obtained as an output from a PSS and is given as an input to an excitation system) and the K_{c-ss} signal (which is obtained as an output from a POD and is given as an input to a TCSC). Examples of outputs from the system are the measured rotor speed (which is given as an input to a PSS) and measured power flow in a line (which is given as an input to a POD). Note that the output from a

Table 2.2 Electromechanical modes with normalized participation factors

Mode	Damping ratio (%)	Frequency (Hz)	State	PF	State	PF
1	2.622	0.398	δ_{15}	1	ω_{15}	1
2	3.482	0.521	ω_{16}	1	δ_{16}	0.999
3	3.502	0.601	ω_{13}	1	δ_{13}	1
4	4.987	0.793	ω_{15}	1	δ_{15}	0.999
5	5.971	0.998	ω_{3}	1	δ_{3}	0.999
6	5.549	1.125	ω_{12}	1	δ_{12}	0.999
7	6.794	1.156	ω_{5}	1	δ_{5}	0.999
8	5.343	1.162	ω_{12}	1	δ_{12}	1
9	7.255	1.213	ω_{2}	1	δ_{2}	0.999
10	3.839	1.291	ω_{10}	1	δ_{10}	1
11	1.076	1.394	ω_{9}	1	δ_{9}	0.806
12	9.521	1.514	ω_{4}	1	δ_{4}	0.998
13	7.3	1.529	ω_{8}	1	δ_{8}	0.999
14	9.209	1.545	ω_{7}	1	δ_{7}	0.999
15	5.969	1.802	ω_{11}	1	δ_{11}	0.999

controller is given as an input to a controllable device in the system, while an output from the system is given as an input to a controller. The steps for the design of a POD controller are described in Section 2.2.3.7, after introducing some more concepts.

2.2.3.5 Electromechanical modes

The modes which have high participation from rotor angles and rotor speeds of various machines are called electromechanical modes. The electromechanical modes with frequencies in the range 0.1 to 0.9 Hz are called interarea modes, while the rest of the electromechanical modes are local machine modes. Table 2.2 shows the electromechanical modes for the 68-bus test system, and it can be inferred from the table that all the electromechanical modes of the test system have a damping ratio of less than 10% (note that this means that the low-frequency interarea modes are poorly damped). The top four modes in Table 2.2 are interarea modes while the rest are local modes. For each mode the highest two participating states are also tabulated in Table 2.2 and arranged in increasing order of normalized participation factors.

2.2.3.6 Interarea modes and mode shapes

The interarea modes are named so because in these modes the participating machines divide into two groups, and the two groups oscillate against each other. If the interarea modes are poorly damped, or unstable, then the two groups may lose synchronism completely and this leads to system breakdown. Also, as the interarea modes have low frequencies as compared to other modes, for a given damping ratio, the corresponding oscillations take much more time to die down than those of the other modes. A 10% or more damping ratio gives an acceptable system performance, and hence control methods are designed to give 10% damping ratio to all the interarea modes (or at least to the most critical interarea mode[s]).

The phenomenon of all the machines dividing into two groups may be better understood by the help of mode shapes. Mode shapes are the polar plots of the eigenvectors of a mode corresponding to the desired states. In Matlab, "*feather*" or "*compass*" functions may be used for plotting the mode shapes.

Fig. 2.8 shows the mode shapes of the interarea modes, in which the eigenvectors (corresponding to each machine's rotor speed) of all the interarea modes are plotted using both *feather* and *compass* functions of MATLAB – the rectangular plots are using the former, while the latter gives the circular plots. The division of machines into two opposing groups is evident in all the interarea modes, especially from the circular plots which clearly show the eigenvector arrows pointing in opposite directions corresponding to the two groups.

2.2.3.7 Residue-based linear control design

In this subsection, the steps for control design of a POD [65] using TCSC are described. The steps for PSS design remain exactly same, the only difference being the choice of the controllable device and the input/output pair, as mentioned in Section 2.2.3.4. The objective of this control design is to damp the most critical interarea mode in the system, which is Mode 1, which has the lowest frequency of the four interarea modes (as given in Table 2.2 or Fig. 2.8).

The first step of control design is to select the input/output pair for the controller. This requires us to select the line power signal which will be obtained as output from the system, to be given as an input to the controller, while the output from the controller or the input to the system is the K_{c-ss} signal to the TCSC. This is done by finding the residue of line power flow for all the 83 lines in the system in Mode 1, with K_{c-ss} as the

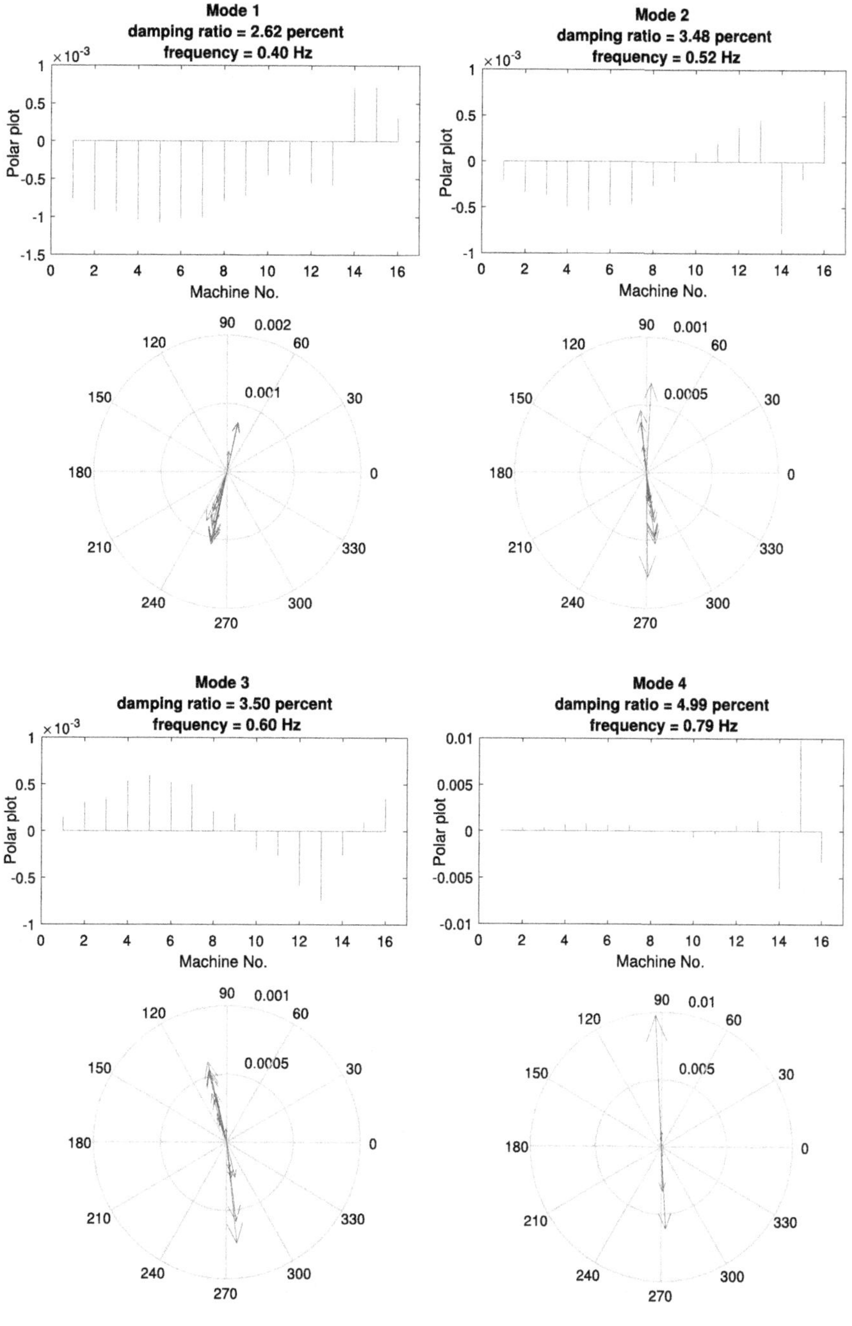

Figure 2.8 Mode shapes for interarea modes.

Table 2.3 Normalized residues of the active line power flows in Mode 1

Mode 1, $\zeta = 0.0206, f = 0.398$ **Hz**

Signal	Residue	Normalized absolute value of residue
P_{51-45}	$0.0427 + j1.070$	1.0000
P_{18-50}	$0.0347 + j1.008$	0.9424
P_{51-50}	$-0.0358 - j1.007$	0.9411
P_{13-17}	$-0.1421 + j0.994$	0.9374
P_{47-53}	$0.0512 + j0.814$	0.7618

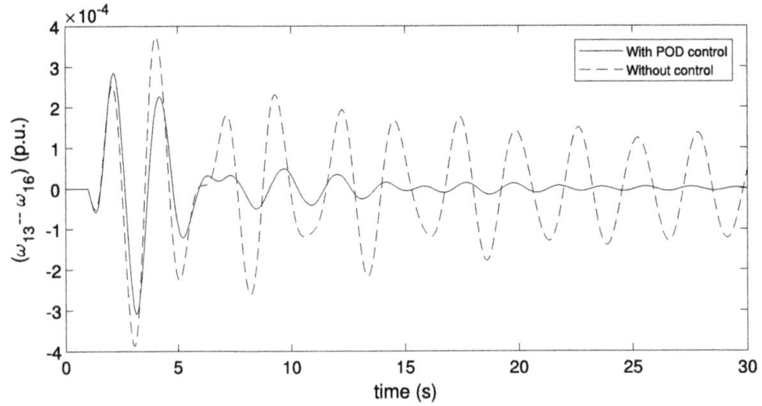

Figure 2.9 Comparison of ω_{13}–ω_{16} for POD control vs. no control.

input. Table 2.3 shows the sorted and normalized absolute residues of the top five active line power signals in Mode 1, with K_{c-ss} as the input.

It can be seen in Table 2.3 that the signals P_{51-45} and P_{18-50} have highest and second-highest residues in Mode 1, with K_{c-ss} as the input. Although P_{51-45} has the highest residue in Mode 1 with K_{c-ss} as the input, it is more convenient to use P_{18-50} as this signal is local to the TCSC and its residue is also not much smaller than that for P_{51-45}. Thus, K_{c-ss} and P_{18-50} are selected as the input/output pair for control design.

The next step is to design the phase compensation part of the controller, denoted by $H_{POD}(s)$. The change of the eigenvalue should be such that it is directed towards the left of the complex plane, so that the total phase is 180 degrees. If the residue of K_{c-ss} and P_{18-50} in Mode 1 is given by R and Mode 1 is denoted by λ, then the phase lead, Φ, required by the controller is

$$\Phi = 180° - \arg\{R\}. \tag{2.55}$$

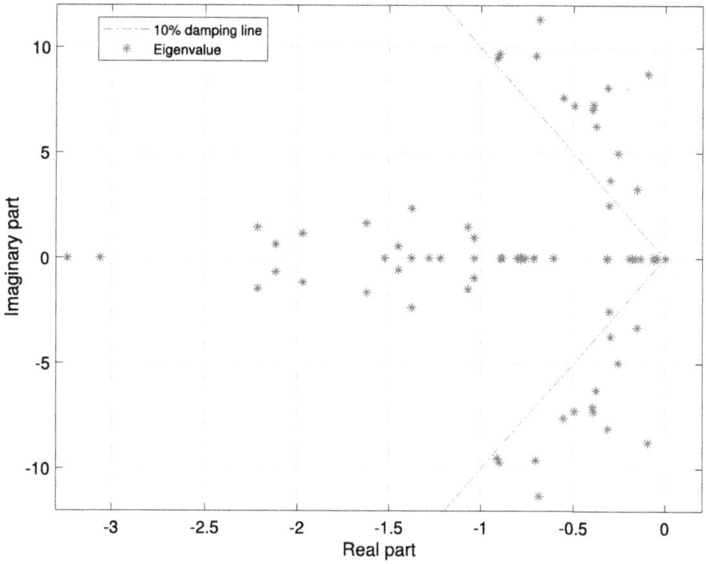

Figure 2.10 Plot of the eigenvalues for the system with POD control.

The phase compensation stages of the controller which can provide this phase lead, provided that c lead–lag stages and a washout filter with time constant T_w are used, are given as follows (a washout filter is usually included in linear controllers to remove any steady-state DC components from the controller output):

$$
\begin{aligned}
H_{POD}(s) &= \frac{sT_w}{1+sT_w}\left[\frac{1+sT_1}{1+sT_2}\right]^c,\\
T_2 &= \left[|\text{imag}\{\lambda\}|\sqrt{\left\{\frac{1-\sin\left(\frac{\Phi}{c}\right)}{1+\sin\left(\frac{\Phi}{c}\right)}\right\}}\right]^{-1},\\
T_1 &= \left\{\frac{1-\sin\left(\frac{\Phi}{c}\right)}{1+\sin\left(\frac{\Phi}{c}\right)}\right\}T_2.
\end{aligned}
\tag{2.56}
$$

The final step is to find the gain of the controller. If the required damping ratio is ζ_r, then the gain, K_{POD}, can be set as follows:

$$
K_{POD} = \text{real}\left\{\frac{-\zeta_r|\lambda| - \text{real}\{\lambda\}}{H_{POD}(\lambda)R}\right\}. \tag{2.57}
$$

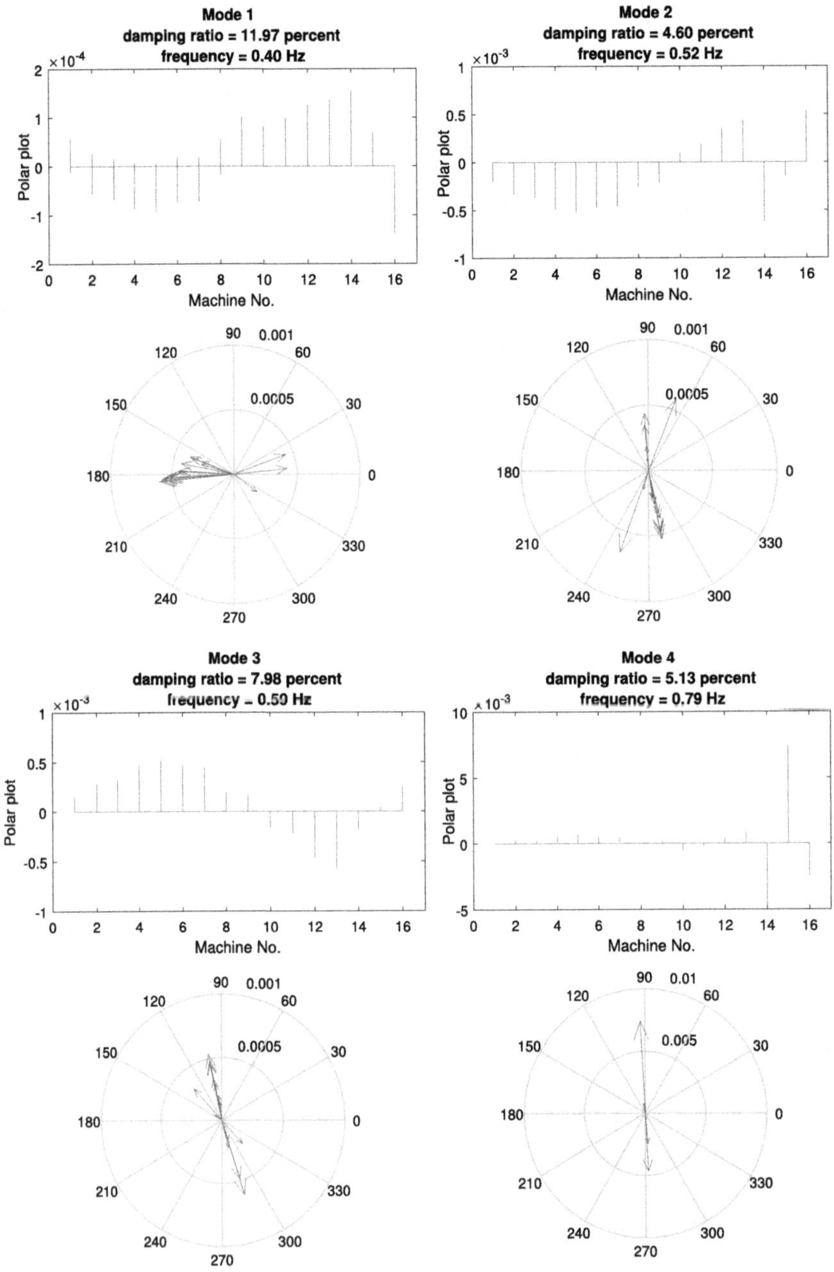

Figure 2.11 Mode shapes for interarea modes for the system with POD control.

Using (2.55)–(2.57), taking $\zeta_r = 0.12$, $T_w = 10$, and $c = 2$, and with $\lambda = -0.0655 + j2.498$ and $R = 0.0347 + j1.008$, the final expression for the POD controller is

$$\begin{aligned} K_{c-ss} &= K_{POD} H_{POD}(s) \Delta P_{18-50} \\ &= -1.4189 \frac{10s}{1+10s} \left[\frac{1+0.162s}{1+0.991s} \right]^2 \Delta P_{18-50}. \end{aligned} \tag{2.58}$$

For assessing the performance of the above controller, simulation and linear analysis were performed for the closed-loop system with the POD controller. The results for time–domain simulation are shown in Fig. 2.9, while those for linear analysis are shown in Fig. 2.10 and Fig. 2.11. It can be observed in Fig. 2.9 that the oscillations get dampened within 30 s. This is because the most critical mode of the system (Mode 1) is now properly damped, with a damping ratio of 11.97%, and the rest of the interarea modes are also better damped than the "no-control" case. This can be clearly seen in Fig. 2.10 and Fig. 2.11. In Fig. 2.10, one of the electromechanical modes (Mode 1) is now inside the 10% damping conic section, while the "no-control" case shown in Fig. 2.7 has all the electromechanical modes outside this conic section. The *feather* plots of Fig. 2.11 clearly show that the four interarea modes have better damping ratios as compared to the "no-control" case shown in Fig. 2.8. Also, the designed control works as expected, as the required damping ratio for Mode 1 for the control design was taken as 0.12 and the obtained damping ratio of 0.1197 is very close.

The above steps for control design can also be used to design controllers for the other interarea modes, and the outputs of all the controllers can then be added and given as an input to the TCSC. One thing to be noted here is that since all the designed controllers modify the same input signal, K_{c-ss}, they can interfere with each other. Therefore, for such controllers, although the steps for the phase compensation part remain the same, deciding the gains of these controllers requires a bit of trial-and-error, or other optimization techniques. In subsequent chapters, other methods of both linear and nonlinear control design will be discussed and studied, which involve minimal use of trial-and-error and provide optimal control laws.

CHAPTER 3

Centralized Dynamic Estimation and Control

Chapter 2 presented the basics of power system modeling. Once the models of various components of a power system are ready, estimation and control of system dynamics can be performed. System dynamics can be controlled using centralized or decentralized schemes. This chapter presents a centralized estimation and control method and discusses its limitations, which pave the way for the decentralized estimation and control techniques discussed later.

A networked control system (NCS) approach utilizing modern communication concepts is very appropriate in the context of centralized dynamic estimation and control of power systems. An NCS is defined as a system in which the control loops are closed through a real-time communication network [66]. NCS enables control execution from long distances by connecting cyberspace to physical space. It has been successfully applied in other technology areas such as space and terrestrial exploration, aerospace, auto-mobiles, factory automation, and industrial process control. NCS offers many advantages over traditional control architectures. Addition of new sensors, actuators, or controllers in traditional control architectures can result in a significant increase in wiring and the complexity of the control system, leading to increased costs and reduced flexibility with every new component. Utilizing a communication network for connecting these components can reduce the complexity of the system and lower the maintenance costs, with nominal investments, as networked controllers allow data to be shared efficiently. Furthermore, networked control offers high flexibility as new control system components can be added with little extra cost and without making significant structural changes to the system. The advantages of NCS over traditional control systems have been elaborated in [66].

Packet switching-based communication networks are the most widely adopted systems for fast, economic, and stable data transfer over both large and small distances through dynamic path allocation. Unlike the traditional circuit switching-based networks, packet switching-based communication networks do not require a dedicated link to be established between the

Dynamic Estimation and Control of Power Systems
https://doi.org/10.1016/B978-0-12-814005-5.00014-5

sending and receiving ends. Circuit switching is not only inefficient and costlier than packet switching, but also the link failure rate increases for large transmission distances, and the failure cannot be dynamically corrected, unlike in packet switching [67]. This is the reason why most of the current research in NCS is based on packet switching technology. However, packet switching-based networks also suffer from some problems, such as packet dropout, network-induced delays, and packet disordering [66]. These factors can possibly degrade the performance of the control of power system dynamics and small signal stability. As explained in Chapter 1, in the context of interconnected power systems, the control of oscillatory stability is very time-critical as uncontrolled oscillations in the past have led to several power blackouts. Therefore, these factors need to be analyzed thoroughly for assessing the suitability of the NCS approach to wide-area control of power systems [22].

Over the past decade, substantial research has been undertaken to model NCS and study the effects of packet dropout and time delays on the control design and the stability of the NCS ([68], [69], [70], and [71]). In the majority of the literature on NCS relating to power systems, it is assumed that the transmission of signals to and from the central control unit occurs over an ideal, lossless, and delay-free communication network. A few exceptions to this are [72], [73], and [74]. In [72], the effect of network-induced time delays has been considered using a wide-area measurement system (WAMS)-based state feedback control methodology. In [73], an estimation of distribution algorithm-based speed control of a networked direct current (DC) motor system has been studied, and in [74] the effect of communication bandwidth constraints on the stability of WAMS-based power system control has been studied. But all these papers have other limitations. For instance, in [72], it is not explained how the various system states (such as the rotor angle, rotor velocity, and transient voltages) are estimated before using them for state feedback, and the power system model considered in the paper is too simplistic to represent actual power system dynamics. In [73], only a local network-based control of a single DC motor system is considered instead of considering the networked control of a complete power system. In [74], the chief problems associated with networked control, which are packet loss and delay, are not considered. This chapter analyzes the effects of packet dropout on the oscillatory stability response of a networked controlled power system (NCPS).

A detailed model for an NCPS is presented in Section 3.1. Section 3.2 presents linear matrix inequality (LMI)-based stability analysis of the devel-

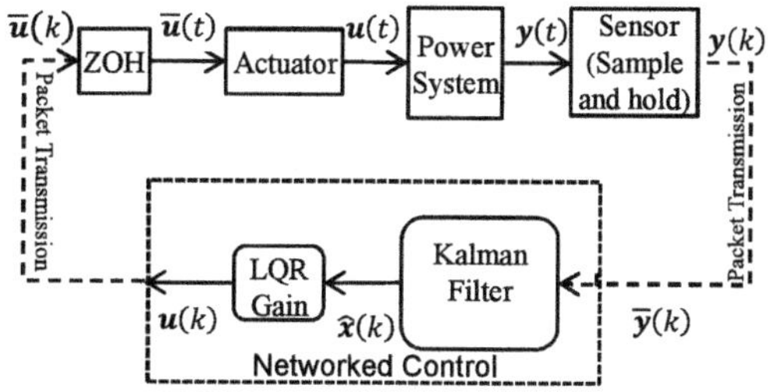

Figure 3.1 A reduced model of the NCPS.

oped NCPS and derives the probability threshold of the packet dropout rate while guaranteeing a specified level of damping of the NCPS. A case study demonstrating the NCPS model is presented in Section 3.3, in which the interarea oscillatory dynamics of the 68-bus test system are controlled using feedback signals transmitted over a communication network. Section 3.4 presents the limitations of the developed NCPS model and Section 3.5 summarizes the chapter.

3.1 NCPS MODELING WITH OUTPUT FEEDBACK

A block diagram of the output feedback-controlled NCPS is shown in Fig. 3.1. The model is described as hybrid continuous–discrete system in which the power system is the continuous, while the networked controller is the discrete part. The NCPS model is hybrid in the sense that the power system is deterministic while the networked controller is stochastic in nature.

In Fig. 3.1, the block "Power system" represents the open-loop power system, the oscillatory dynamics of which need to be controlled. To this effect, the real-power deviations in some of the lines are measured in real-time using current transformers (CTs) and potential transformers (PTs) [75] and represented by $\boldsymbol{y}(t)$ in the block diagram. These are then sampled at the sampling rate of the communication network using digital devices such as phasor measurement units (PMUs) and intelligent electronic devices (IEDs) and then sent over the communication network as discrete data packets, $\boldsymbol{y}(k)$. User datagram protocol (UDP) is used for packet transmission, and packet loss occurs during transmission. The final data which are received at the control unit after packet loss are given by $\bar{\boldsymbol{y}}(k)$. The control unit consists of a linear quadratic Gaussian (LQG) controller, which is a combination of a Kalman filter and a linear quadratic regulator (LQR). The Kalman filter requires a linearized, discretized, and reduced power system model and the output data packets arriving at the controller, $\bar{\boldsymbol{y}}(k)$, to estimate the states, $\hat{\boldsymbol{x}}(k)$. The state estimates are then multiplied by the LQR gain to produce the control signals $\boldsymbol{u}(k)$, which are then sent over the communication network to the actuators. The packets arrive at discrete to analogue converters (DACs), which are zero-order-hold (ZOH) devices and convert the discrete control signals after packet loss, $\bar{\boldsymbol{u}}(k)$, into continuous control signals, $\bar{\boldsymbol{u}}(t)$. These continuous signals control the actuators, which are the flexible alternating current (AC) transmission system (FACTS) devices, more commonly known as FACTS controllers. The inputs $\boldsymbol{u}(t)$ to the power system are the percentage compensations provided by the FACTS controllers to control the power flow in the lines on which the FACTS controllers are installed. All the variables in the model have been expressed in p.u., except the time variables which are expressed in s. A detailed description of each component of the NCPS model is presented as follows.

3.1.1 State space representation of power system

The state space representation of a power system is obtained through linearization of the DAEs (described in Chapter 2) around an initial operating point. The order of the system is reduced to speed up the controller design algorithm and also to reduce the order of the controller. On applying balanced model reduction based on singular value decomposition, as given in [76], only the unstable and/or poorly damped electromechanical modes of the power system are retained in the reduced model. In MATLAB, balanced model reduction is performed using the "*balred*" function. Let the reduced-order model be written as follows:

$$\Delta\dot{\boldsymbol{x}}(t) = \boldsymbol{A}_R \Delta\boldsymbol{x}(t) + \boldsymbol{B}_R \Delta\boldsymbol{u}(t), \tag{3.1}$$

$$\Delta\boldsymbol{y}(t) = \boldsymbol{C}_R \Delta\boldsymbol{x}(t), \tag{3.2}$$

where $\boldsymbol{A}_R \in \mathbb{R}^{m\times m}$, $\boldsymbol{B}_R \in \mathbb{R}^{m\times p}$, and $\boldsymbol{C}_R \in \mathbb{R}^{q\times m}$ are the reduced state space matrices and $\boldsymbol{x} \in \mathbb{R}^m$, $\boldsymbol{u} \in \mathbb{R}^p$, and $\boldsymbol{y} \in \mathbb{R}^q$ are the vectors of state variables, inputs, and outputs, respectively. It should be noted that after balanced reduction of the full model, only the state variables and the state matrices get reduced in order; the inputs $\boldsymbol{u}$ and the outputs $\boldsymbol{y}$ remain the same as in the original full model. Also, out of the various possible measurable outputs (which are the line powers in the context of NCPS), only those outputs are selected in $\boldsymbol{y}$ which have high observability of the unstable and/or poorly damped electromechanical modes of the power system.

3.1.2 Sensors and actuators

The sensors (in the context of NCPS, they are CTs, PTs, and PMUs) send the feedback signals to the controller over the communication network at a regular interval of T_0, which is the sampling period of the communication network. The DACs convert the discrete control signals after packet loss into continuous control signals. The DACs are event-driven ZOH devices, each one of which holds the input to the power system in a given cycle. In the next cycle it holds its previous value if there is no new input due to packet drop, otherwise it holds the new input. The outputs of the DACs control the FACTS controllers, which are the actuators; the inputs $\boldsymbol{u}(t)$ to the power system are the percentage compensations provided by the FACTS controllers. For the $(k+1)$th time cycle, (3.1) reduces to

$$\Delta\dot{\boldsymbol{x}}(t) = \boldsymbol{A}_R\Delta\boldsymbol{x}(t) + \boldsymbol{B}_R\Delta\boldsymbol{u}(kT_0);\, 0 \le t - kT_0 < T_0. \tag{3.3}$$

Solving (3.3) with initial condition $(\Delta\boldsymbol{x}(kT_0), \Delta\boldsymbol{u}(kT_0))$ and a constant input $\Delta\boldsymbol{u}(kT_0)$ [77], we get

$$\Delta\boldsymbol{x}((k+1)\,T_0) = \boldsymbol{A}\Delta\boldsymbol{x}(kT_0) + \boldsymbol{B}\Delta\boldsymbol{u}(kT_0), \tag{3.4}$$

$$\boldsymbol{A} = e^{\boldsymbol{A}_R T_0};\, \boldsymbol{B} = \boldsymbol{A}_R^{-1}\left(e^{\boldsymbol{A}_R T_0} - \boldsymbol{I}\right)\boldsymbol{B}_R. \tag{3.5}$$

Denoting $\Delta\boldsymbol{x}(kT_0)$ as $\boldsymbol{x}_k$, $\Delta\boldsymbol{u}(kT_0)$ as $\bar{\boldsymbol{u}}_k$ (where $\bar{\boldsymbol{u}}_k$ is the uncertain input after packet dropout), $\Delta\boldsymbol{y}(kT_0)$ as $\boldsymbol{y}_k$, and $\boldsymbol{C}_R$ as $\boldsymbol{C}$ and also including a white Gaussian measurement noise $\boldsymbol{v}_k$ and a white Gaussian process noise $\boldsymbol{w}_k$ in the model, we get

$$\boldsymbol{x}_{k+1} = \boldsymbol{A}\boldsymbol{x}_k + \boldsymbol{B}\bar{\boldsymbol{u}}_k + \boldsymbol{w}_k;\; \boldsymbol{y}_k = \boldsymbol{C}\boldsymbol{x}_k + \boldsymbol{v}_k. \tag{3.6}$$

3.1.3 Communication protocol, packet delay, and packet dropout

In the model design process two classes of communication protocols have been considered. In transmission control protocol (TCP)-like protocols the acknowledgments that the receiver received the packets are sent back to the sender, while in user datagram protocol (UDP)-like protocols they are not sent. In the TCP-like case, unlike in the UDP-like case, the lost packets can be resent because of the availability of the acknowledgments. So the separation principle, as explained in [78], holds only in the case of TCP-like protocols, and hence the controller and the estimator can be designed

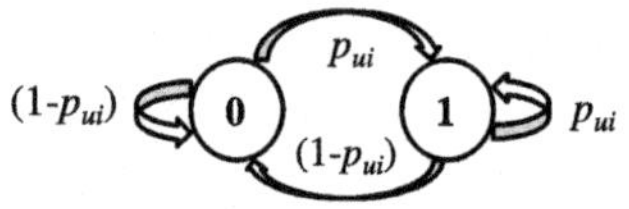

Figure 3.2 Markov chain for the *i*th input channel's delivery indication.

independently [79]. In the UDP-like case no known optimal regulator exists and one can design a suboptimal solution based on a Kalman-like estimator and a LQG-like state feedback controller, as shown in Fig. 3.1. Although UDP-like protocol results in a suboptimal solution, it is preferred over a TCP-like protocol as it may be extremely difficult to both analyze and implement a TCP-like control scheme [79]. Hence, a UDP-like scheme is used in the NCPS model. The time delays and dropouts of packets are modeled such that a packet is assumed to be lost, unless its time delay is less than the sampling interval of the system. This fact is one of the factors while deciding the sampling duration, the other factor being the type of control needed, as explained in Section 3.3.2.1. If a packet is lost, the output of the receiver is held at the last successfully received packet.

The packet loss over the network usually follows a random process. In the present analysis an independent Bernoulli process has been used to model the packet loss [79]. The input $\bar{\boldsymbol{u}}_k$ at the actuator and $\bar{\boldsymbol{y}}_k$ at the estimator are modeled as

$$\bar{\boldsymbol{u}}_k = \boldsymbol{\alpha}_k \boldsymbol{u}_k, \ \bar{\boldsymbol{y}}_k = \boldsymbol{\beta}_k \boldsymbol{y}_k, \tag{3.7}$$

where $\boldsymbol{\alpha}_k = diag\left(\alpha_k^1, \alpha_k^2, ..., \alpha_k^p\right)$ is a stationary diagonal binary random matrix, in which the value of α_k^i is equal to one with a probability p_{ui}, indicating that the *i*th component of $\boldsymbol{u}_k$ is delivered; while its value is equal to zero with a probability $(1 - p_{ui})$, indicating that the component is lost (Fig. 3.2). Similarly, $\boldsymbol{\beta}_k = diag\left(\beta_k^1, \beta_k^2, ..., \beta_k^q\right)$ is the stationary diagonal binary random matrix for the delivery indication of $\boldsymbol{y}_k$. Also, p_{ui} is termed the packet delivery probability (PDP) of the *i*th input channel, while p_{yi} is the PDP of the *i*th output channel.

Remark. The assumption of an independent Bernoulli packet loss model is not valid when the communication channel is congested. In a congested channel the packet loss occurs in bursts and follows a two-state Markov chain model, also known as Gilbert model [80]. Fig. 3.3 shows this model, where "1" represents the state of packet delivery and "0" represents the

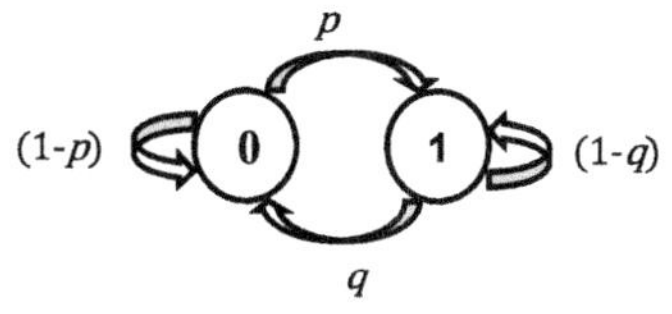

Figure 3.3 Markov chain for Gilbert process.

state of packet loss; and the probability of transition from state "0" to state "1" is p and the probability of transition from state "1" to state "0" is q. When p is equal to $(1 - q)$, this model reduces to the Bernoulli model.

The drawback of using the Gilbert model in stability analysis is that this model is not a memoryless model, which means that the probability of packet delivery depends on the current channel state, and it fluctuates between p and $(1 - q)$ depending on whether the current channel state is "0" or "1", respectively. Mathematical representation of such a fluctuating probability of packet delivery becomes practically infeasible. A practical alternative for approximating the Gilbert model with the Bernoulli model can be to set the communication channel's probability of packet delivery as p if $p < (1 - q)$ and as $(1 - q)$ if $(1 - q) \leq p$. Thus, the approximated Bernoulli model represents the worst-case scenario of packet delivery performance given by the Gilbert model, as the smaller probability of the two possible packet delivery probabilities from the Gilbert model is assumed to be the constant PDP in the approximated Bernoulli model.

3.1.4 Controller

For an open-loop LTI system given by (3.6), with input $\bar{\boldsymbol{u}}_k$ defined by (3.7), the quadratic cost function $\boldsymbol{J}$ is given by

$$\boldsymbol{J} = \frac{1}{N}\mathbb{E}\left\{\boldsymbol{x}_N^T \boldsymbol{Q}\boldsymbol{x}_N + \sum_{k=1}^{N-1}\left[\boldsymbol{x}_k^T \boldsymbol{Q}\boldsymbol{x}_k + \boldsymbol{u}_k^T \boldsymbol{\alpha}_k \boldsymbol{R}\boldsymbol{\alpha}_k \boldsymbol{u}_k\right]\right\}, \tag{3.8}$$

where N is the number of samples, $\mathbb{E}$ is the expectation value, T denotes the transpose of a vector or a matrix, $\boldsymbol{Q}$ is a positive definite matrix denoting state costs, $\boldsymbol{R}$ is a positive semidefinite matrix denoting input costs, and it is assumed that the full state information of the LTI system is available (we get this information from the state estimator). Minimizing $\boldsymbol{J}$ with respect to $\boldsymbol{u}_k$ results in the following Riccati-like difference equation, as explained in [81]:

$$\begin{aligned} \boldsymbol{M}_{k+1} = & \boldsymbol{A}^T\boldsymbol{M}_k\boldsymbol{A} + \mathbf{Q} \\ & - \boldsymbol{A}^T\boldsymbol{M}_k\boldsymbol{B}\mathbb{E}[\boldsymbol{\alpha}](\boldsymbol{R} + \mathbb{E}[\boldsymbol{\alpha}\boldsymbol{B}^T\boldsymbol{M}_k\boldsymbol{B}\boldsymbol{\alpha}])^{-1}\mathbb{E}[\boldsymbol{\alpha}]\boldsymbol{B}^T\boldsymbol{M}_k\boldsymbol{A}, \end{aligned} \tag{3.9}$$

where $\boldsymbol{M}_0$ is $\mathbf{Q}$ and $\mathbb{E}[\boldsymbol{\alpha}]$ is the expectation value of $\boldsymbol{\alpha}_k$ (the subscript k is removed in $\mathbb{E}[\boldsymbol{\alpha}]$ as $\boldsymbol{\alpha}_k$ is stationary). If we obtain a steady-state solution $\boldsymbol{M} = \boldsymbol{M}_\infty$ for (3.9) as $k \to \infty$, then the LTI open-loop system is infinite-horizon stabilizable in mean-square sense, provided the pair $(\boldsymbol{A}, \boldsymbol{B})$ is controllable; the pair $(\boldsymbol{A}, \mathbf{Q}^{1/2})$ is observable, where $\mathbf{Q} = (\mathbf{Q}^{1/2})^T\mathbf{Q}^{1/2}$. The infinite-horizon control policy for such a system is a state feedback policy, given by

$$\boldsymbol{u}_k = \boldsymbol{L}\widehat{\boldsymbol{x}}_k; \ \boldsymbol{L} = -(\boldsymbol{R} + \mathbb{E}[\boldsymbol{\alpha}\boldsymbol{B}^T\boldsymbol{M}\boldsymbol{B}\boldsymbol{\alpha}])^{-1}\mathbb{E}[\boldsymbol{\alpha}]\boldsymbol{B}^T\boldsymbol{M}\boldsymbol{A}, \tag{3.10}$$

where $\widehat{\boldsymbol{x}}_k$ is the estimated state and $\boldsymbol{L}$ is the LQG gain.

3.1.5 Estimator

The controller uses the output from the state estimator to generate the control command which is sent over the network to the actuator in the power system. The estimator uses the information vector, which consists of the control command and the intermittent plant output delivered to the estimator via the network, to generate a best estimate of the state of the system. It was shown in [82] that even in the case of intermittent observations, the Kalman filter is still the best linear estimator for LTI systems with stationary Gaussian noise processes, provided that only time is updated when a measurement packet is dropped. When a measurement is received, both the time and measurement are updated. The Kalman filtering equations for such an LTI system are given as follows.

3.1.5.1 State prediction step

$$\hat{\boldsymbol{x}}_k^- = \boldsymbol{A}'\hat{\boldsymbol{x}}_{k-1}, \ \boldsymbol{A}' = (\boldsymbol{A} + \boldsymbol{B}\mathbb{E}[\boldsymbol{\alpha}]\boldsymbol{L}), \tag{3.11}$$

$$\boldsymbol{P}_{xk}^- = \boldsymbol{A}'\boldsymbol{P}_{x(k-1)}\boldsymbol{A}'^T + \boldsymbol{P}_{vk}. \tag{3.12}$$

3.1.5.2 Measurement prediction and Kalman update step

$$\hat{\boldsymbol{x}}_k = \hat{\boldsymbol{x}}_k^- + \boldsymbol{K}_k\boldsymbol{\beta}_k(\boldsymbol{y}_k - \boldsymbol{C}\hat{\boldsymbol{x}}_k^-), \tag{3.13}$$

$$\boldsymbol{P}_{xk} = \boldsymbol{P}_{xk}^- - \boldsymbol{K}_k\boldsymbol{\beta}_k\boldsymbol{C}\boldsymbol{P}_{xk}^-, \tag{3.14}$$

$$\boldsymbol{K}_k = \boldsymbol{P}_{xk}^{-}\boldsymbol{C}^T[\boldsymbol{C}\boldsymbol{P}_{xk}^{-}\boldsymbol{C}^T + \boldsymbol{P}_{wk}]^{-1}. \tag{3.15}$$

For the kth sample, $\hat{\boldsymbol{x}}_k^-$ is the estimated mean of the predicted states, $\boldsymbol{P}_{xk}^-$ is the predicted-state covariance matrix, $\hat{\boldsymbol{x}}_k$ is the estimated mean of the states, $\boldsymbol{P}_{xk}$ is the state covariance matrix, $\boldsymbol{P}_{wk}$ is the covariance matrix of $\boldsymbol{w}_k$, $\boldsymbol{P}_{vk}$ is the covariance matrix of $\boldsymbol{v}_k$, and $\boldsymbol{K}_k$ is the Kalman gain. The equations are valid if and only if $(\boldsymbol{A}', \boldsymbol{C})$ is observable and $(\boldsymbol{A}', \boldsymbol{P}_{vk}^{1/2})$ is controllable. In (3.11) the estimator takes the closed-loop state space matrix $\boldsymbol{A}'$ as $(\boldsymbol{A} + \boldsymbol{B}\mathbb{E}[\boldsymbol{\alpha}]\boldsymbol{L})$ as it can at best have an estimate of the packet dropout rate of the network because it does not receive the acknowledgments of the control packets it sends out to the power system.

3.2 CLOSED-LOOP STABILITY AND DAMPING RESPONSE

The closed-loop model of the NCPS can be summarized as follows, using (3.6)–(3.15):

$$\boldsymbol{x}_{k+1} = \boldsymbol{A}\boldsymbol{x}_k + \boldsymbol{B}\boldsymbol{\alpha}_k\boldsymbol{L}\widehat{\boldsymbol{x}}_k + \boldsymbol{w}_k, \tag{3.16}$$

$$\widehat{\boldsymbol{x}}_{k+1} = \boldsymbol{A}'\widehat{\boldsymbol{x}}_k + \boldsymbol{K}_{k+1}\boldsymbol{\beta}_{k+1}(\boldsymbol{y}_{k+1} - \boldsymbol{C}\boldsymbol{A}'\widehat{\boldsymbol{x}}_k), \tag{3.17}$$

$$\boldsymbol{y}_{k+1} = \boldsymbol{C}(\boldsymbol{A}\boldsymbol{x}_k + \boldsymbol{B}\boldsymbol{\alpha}_k\boldsymbol{L}\widehat{\boldsymbol{x}}_k + \boldsymbol{w}_k) + \boldsymbol{v}_{k+1}. \tag{3.18}$$

A steady-state solution for $\boldsymbol{P}_{xk}$ in (3.14), and hence for $\boldsymbol{K}_k$, may or may not exist for given $\boldsymbol{\alpha}_k$ and $\boldsymbol{\beta}_k$, even if the conditions for the existence of a steady-state solution for a standard Kalman filter hold; but a steady state estimate $\boldsymbol{K} = \mathbb{E}[\boldsymbol{K}_\infty]$ for the Kalman gain may be obtained by iteratively solving (3.12), (3.14), and (3.15) after substituting $\boldsymbol{\beta}_k$ with its expected value $\mathbb{E}[\boldsymbol{\beta}]$. This is the suboptimal Kalman gain which is used for deriving the condition for mean-square stability and adequate damping of the developed NCPS. Writing (3.16)–(3.18) in composite form, after replacing $\boldsymbol{K}_{k+1}$ with its steady-state estimate $\boldsymbol{K}$, we get

$$\begin{bmatrix} \boldsymbol{x}_{k+1} \\ \widehat{\boldsymbol{x}}_{k+1} \end{bmatrix} = \begin{bmatrix} \boldsymbol{I}_m \\ \boldsymbol{K}\boldsymbol{\beta}_{k+1}\boldsymbol{C} \end{bmatrix} \boldsymbol{w}_k + \begin{bmatrix} 0_{m\times q} \\ \boldsymbol{K}\boldsymbol{\beta}_{k+1} \end{bmatrix} \boldsymbol{v}_{k+1}$$
$$+ \underbrace{\begin{bmatrix} \boldsymbol{A} & \boldsymbol{B}\boldsymbol{\alpha}_k\boldsymbol{L} \\ \boldsymbol{K}\boldsymbol{\beta}_{k+1}\boldsymbol{C}\boldsymbol{A} & \boldsymbol{A}' + \boldsymbol{K}\boldsymbol{\beta}_{k+1}\boldsymbol{C}(\boldsymbol{B}\boldsymbol{\alpha}_k\boldsymbol{L} - \boldsymbol{A}') \end{bmatrix}}_{\mathbb{A}(\boldsymbol{\alpha}_k,\, \boldsymbol{\beta}_{k+1})} \begin{bmatrix} \boldsymbol{x}_k \\ \widehat{\boldsymbol{x}}_k \end{bmatrix}. \tag{3.19}$$

The presence of α_k and β_{k+1} in (3.19) makes it a jump linear system (JLS): a system whose state matrices vary randomly with α_k and β_{k+1}. The framework of a JLS and its stability analysis are described in [83] and [84]. A brief overview of the criterion for the stability and the damping in mean-square sense of the NCPS is presented in the next section.

3.2.1 Stability analysis framework of a jump linear system

Let $\boldsymbol{S}_i$ be a set of all the subsets of $\{1, 2, 3, ..., i\}$. Let $\boldsymbol{r} \in \boldsymbol{S}_p$ be a set of indices of all those input delivery indicators whose values are one, i.e. $\boldsymbol{r} = \{i$, such that (s.t.) $\alpha_k^i = 1\}$. For example, for a two-input system ($p = 2$), $\boldsymbol{r}$ can be $\emptyset$ (both the inputs failed to deliver), $\{1\}$ (only 1st input delivered), $\{2\}$ (only 2nd input delivered), or $\{1, 2\}$ (both the inputs delivered). Similarly, let $\boldsymbol{s} \in \boldsymbol{S}_q$ be a set of indices of successful output delivery indicators. As each input delivery indicator α_k^i has two modes (0 or 1) and $\alpha_k^i s$ are p in total, $\boldsymbol{\alpha}_k$ has 2^p modes. Any mode of $\boldsymbol{\alpha}_k$ is expressed as $\boldsymbol{T}_p(\boldsymbol{r})$, $\boldsymbol{r} \in \boldsymbol{S}_p$, where $\boldsymbol{T}_p(\boldsymbol{r})$ is a $p \times p$ diagonal matrix whose (i, i)th element is 1 if $i \in \boldsymbol{r}$, or else it is 0. Similarly, $\boldsymbol{\beta}_{k+1}$ has 2^q modes, and any mode is expressed as $\boldsymbol{T}_q(\boldsymbol{s})$, $\boldsymbol{s} \in \boldsymbol{S}_q$, where $\boldsymbol{T}_q(\boldsymbol{s})$ is a $q \times q$ diagonal matrix whose (i, i)th element is 1 if $i \in \boldsymbol{s}$, or else it is 0. The probability distributions of $\boldsymbol{T}_p(\boldsymbol{r})$, $\boldsymbol{r} \in \boldsymbol{S}_p$, and $\boldsymbol{T}_q(\boldsymbol{s})$, $\boldsymbol{s} \in \boldsymbol{S}_q$, are given by

$$\mathcal{P}_p(\boldsymbol{r}) = \mathbf{P}\left[\boldsymbol{\alpha}_k = \boldsymbol{T}_p(\boldsymbol{r})\right] = \prod_{i \in r} p_{ui} \prod_{i \notin r} 1 - p_{ui}, \tag{3.20}$$

$$\mathcal{P}_q(\boldsymbol{s}) = \mathbf{P}\left[\boldsymbol{\beta}_{k+1} = \boldsymbol{T}_q(\boldsymbol{s})\right] = \prod_{i \in s} p_{yi} \prod_{i \notin s} 1 - p_{yi}, \tag{3.21}$$

where $\mathcal{P}_p(\boldsymbol{r})$ is the resultant probability of data delivery for any combination of input to the plant by the channel characterized by $\boldsymbol{T}_p(\boldsymbol{r})$. Similarly $\mathcal{P}_q(\boldsymbol{s})$ is the resultant probability of data delivery for any combination of plant output channel mode characterized by $\boldsymbol{T}_q(\boldsymbol{s})$.

As $\mathbb{A}(\boldsymbol{\alpha}_k, \boldsymbol{\beta}_{k+1})$ in (3.19) is a function of $\boldsymbol{\alpha}_k$ and $\boldsymbol{\beta}_{k+1}$, it may be reexpressed as $\mathcal{A}(\boldsymbol{r}, \boldsymbol{s})$ in (3.22), i.e.,

$$\mathcal{A}(\boldsymbol{r}, \boldsymbol{s}) = \begin{bmatrix} \boldsymbol{A} & \boldsymbol{B}\boldsymbol{T}_p(\boldsymbol{r})\boldsymbol{L} \\ \boldsymbol{K}\boldsymbol{T}_q(\boldsymbol{s})\boldsymbol{C}\boldsymbol{A} & \boldsymbol{A}' + \boldsymbol{K}\boldsymbol{T}_q(\boldsymbol{s})\boldsymbol{C}(\boldsymbol{B}\boldsymbol{T}_p(\boldsymbol{r})\boldsymbol{L} - \boldsymbol{A}') \end{bmatrix}. \tag{3.22}$$

As the value of $\mathcal{A}(\boldsymbol{r}, \boldsymbol{s})$ depends on the values of $\boldsymbol{r}$ and $\boldsymbol{s}$, it can take any value in a given sample out of the possible 2^{p+q} values, with a corresponding overall probability distribution $\mathcal{P}_p(\boldsymbol{r})\mathcal{P}_q(\boldsymbol{s})$. The NCPS in (3.19) is said to

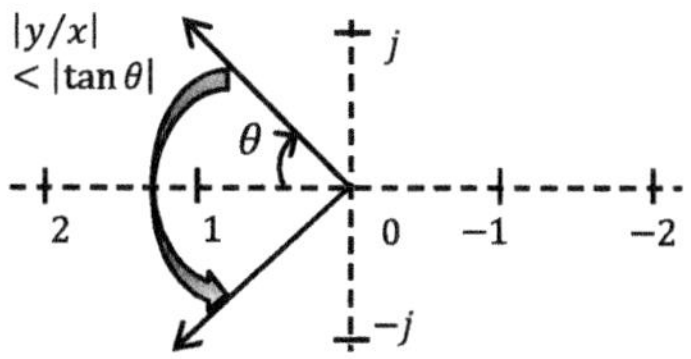

Figure 3.4 $\mathcal{D}$-stability region for damping control of a continuous system.

be mean-square stable if $\lim_{k\to\infty}\mathbb{E}\left\|\begin{bmatrix} x_k \\ \hat{x}_k \end{bmatrix}\right\|^2 = 0$, starting with any state $\begin{bmatrix} x_0 \\ \hat{x}_0 \end{bmatrix}$. Mean-square stability and the damping response of (3.19) can be checked using the following inequalities given in Subsections 3.2.1.1–3.2.1.2. These inequalities are formed using linear combinations of symmetric matrices, and are known as linear matrix inequalities (LMIs).

3.2.1.1 LMIs for mean-square stability

The criterion for the stability of discrete-time JLSs in [83] is applied to obtain the condition for the mean-square stability of the NCPS in (3.22). The satisfaction of the criterion requires the existence of positive definite matrices $\boldsymbol{P}_{r,s}$, $\forall \boldsymbol{r} \in \boldsymbol{S}_p$, $\forall \boldsymbol{s} \in \boldsymbol{S}_q$, such that

$$\boldsymbol{P}_{r,s} > \mathcal{A}(\boldsymbol{r},\boldsymbol{s})\left(\mathcal{P}_p(\boldsymbol{r})\mathcal{P}_q(\boldsymbol{s})\sum_{\boldsymbol{r}'\in\boldsymbol{S}_p,\boldsymbol{s}'\in\boldsymbol{S}_q}\boldsymbol{P}_{r',s'}\right)\mathcal{A}(\boldsymbol{r},\boldsymbol{s})^T. \tag{3.23}$$

There are in total 2^{p+q} LMIs in (3.23).

3.2.1.2 LMIs for adequate damping response

The concept of $\mathcal{D}$-stability [85] has been used to study the adequate damping response of the developed NCPS. This is very practical and useful in the context of power oscillation damping (POD). If $\mathcal{D}$ is a subregion of the complex left half plane and all the closed-loop poles of a dynamical system $\dot{x} = \boldsymbol{A}x$ lie in $\mathcal{D}$, then the system and its state transition matrix $\boldsymbol{A}$ are called $\mathcal{D}$-stable. When $\mathcal{D}$ is the entire left half plane, then $\mathcal{D}$-stability reduces to asymptotic stability. For damping control analysis, the $\mathcal{D}$-region of interest is $\mathcal{D}(\Theta)$ of complex numbers $(x+jy)$ s.t. $|y/x| < |\tan\Theta|$ (Fig. 3.4). Thus a specified and required damping of interarea modes becomes an important criterion for NCS design and analysis.

Confining the closed-loop poles of the system to the region shown in Fig. 3.4 ensures a minimum damping ratio $\zeta_0 = \cos\Theta$. This in turn bounds the decay rate and the settling time for the corresponding oscillatory interarea modes of the system. Power systems usually require an operating constraint that all the disturbances in the system should settle to less than a fixed percent (usually 15%) of the maximum overshoot within a few seconds (usually 10–15 s) of the start of the disturbance to the system [7]. As the interarea modes usually lie in the frequency range 0.2–1.0 Hz, they have longer settling times and lower decay rates than other modes. In this chapter, the margin for $\mathcal{D}$-stability is taken as a minimum damping ratio of 0.1 for all the closed-loop interarea modes, as a damping ratio of 0.1 corresponds to a settling time of 15 s for a modal frequency of 0.2 Hz.

Lemma 3.1. *The closed-loop NCPS in* (3.19) *is expected to have all of its equivalent continuous-time poles with damping ratios* $\zeta > \cos\Theta$ *if and only if there exist positive definite matrices* $\mathbf{Q}_{r,s}$, $\forall \boldsymbol{r} \in \boldsymbol{S}_p$, $\forall \boldsymbol{s} \in \boldsymbol{S}_q$, *such that*

$$\begin{aligned}&(\boldsymbol{W} \otimes \mathcal{A}_c(\boldsymbol{r}, \boldsymbol{s}))\mathbf{Q}_{r,s} + \mathbf{Q}_{r,s}(\boldsymbol{W} \otimes \mathcal{A}_c(\boldsymbol{r}, \boldsymbol{s}))^T \\ &\quad + \left(\mathcal{P}_p(\boldsymbol{r})\mathcal{P}_q(\boldsymbol{s}) - 1\right)\mathbf{Q}_{r,s} + \mathcal{P}_p(\boldsymbol{r})\mathcal{P}_q(\boldsymbol{s})\Big(\sum_{r' \in S_p, s' \in S_q, r' \neq r, s' \neq s} \mathbf{Q}_{r',s'}\Big) < 0,\end{aligned} \tag{3.24}$$

$$\text{where } \mathcal{A}_c(\boldsymbol{r}, \boldsymbol{s}) = \ell n(\mathcal{A}(\boldsymbol{r}, \boldsymbol{s}))/T_0,\ \boldsymbol{W} = \begin{bmatrix} \sin\Theta & \cos\Theta \\ -\cos\Theta & \sin\Theta \end{bmatrix}, \tag{3.25}$$

ℓn is the natural logarithm of a square matrix, and T_0 is the sampling period of the NCPS.

Proof. The damping region of interest $\mathcal{D}(\Theta)$ shown in Fig. 3.4 is applicable only to the continuous-time representation of a dynamic system; the discrete-time equivalent of this region is a logarithmic spiral and is very difficult to represent using matrices and LMIs. We therefore consider the continuous-time equivalent of $\mathcal{A}(\boldsymbol{r}, \boldsymbol{s})$, which is given by (3.24) as $\mathcal{A}_c(\boldsymbol{r}, \boldsymbol{s})$. Using [86] we know that a dynamic system $\dot{\boldsymbol{x}} = \boldsymbol{A}\boldsymbol{x}$ is $\mathcal{D}$-stable in the region $\mathcal{D}(\Theta)$ if and only if $\boldsymbol{W} \otimes \boldsymbol{A}$ is asymptotically stable. This holds because the eigenvalues of $\boldsymbol{W}$ are $e^{\pm j(\frac{\pi}{2} - \Theta)}$. The eigenvalues of the Kronecker product of two matrices are the product of the eigenvalues of individual matrices. Hence, the eigenvalues of $\boldsymbol{W} \otimes \boldsymbol{A}$ are two sets of eigenvalues of $\boldsymbol{A}$, one set rotated by an angle $(\frac{\pi}{2} - \Theta)$ and another one by $-(\frac{\pi}{2} - \Theta)$. All those eigenvalues of $\boldsymbol{A}$ which lie outside $\mathcal{D}(\Theta)$ get rotated into the right half plane in $\boldsymbol{W} \otimes \boldsymbol{A}$, and hence $\boldsymbol{W} \otimes \boldsymbol{A}$ is asymptotically stable if and only if none of the eigenvalues of $\boldsymbol{A}$ lie outside $\mathcal{D}(\Theta)$, i.e., if and only if $\boldsymbol{A}$ is adequately damped. So, the asymptotic stability of $\boldsymbol{W} \otimes \mathcal{A}_c(\boldsymbol{r}, \boldsymbol{s})$ implies $\mathcal{D}$-stability of

$\mathcal{A}_c(\boldsymbol{r}, \boldsymbol{s})$. As $\mathcal{A}_c(\boldsymbol{r}, \boldsymbol{s})$ is a jump-linear mode of the stochastic system in (3.19) with a modal probability of $\mathcal{P}_p(\boldsymbol{r})\mathcal{P}_q(\boldsymbol{s})$, the matrix $\boldsymbol{W} \otimes \mathcal{A}_c(\boldsymbol{r}, \boldsymbol{s})$ is also a modal matrix of the same probability as $\mathcal{A}_c(\boldsymbol{r}, \boldsymbol{s})$; and the mean-square stability of $\boldsymbol{W} \otimes \mathcal{A}_c(\boldsymbol{r}, \boldsymbol{s})$ [83] (given by (3.24)) implies the $\mathcal{D}$-stability of $\mathcal{A}_c(\boldsymbol{r}, \boldsymbol{s})$ in a mean-square sense, i.e., its electromechanical modes are expected to have $\zeta > \cos\Theta$. □

Remark. It should be noted that an additional pole placement constraint which is desired (besides the constraint of a minimum damping ratio) is that the real part of each mode should be less than a specified minimum value (usually -0.1), so that none of the modes are very close to the imaginary axis. This constraint is relevant to the modes which have very small modal frequencies (in the range of 0.0–0.16 Hz) as the real parts of only these modes can be greater than -0.1 even if they satisfy the constraint of a minimum damping ratio of 0.1. Thus, this additional constraint has been relaxed in the aforementioned stability analysis, as the analysis focuses on interarea modes (which have modal frequencies greater than 0.2 Hz). Another reason for relaxing the constraint is that it is mathematically difficult to include an additional constraint in the above lemma.

Remark. If all the channels have the same characteristics, their PDPs become equal to each other ($p_{ui} = p_{yi} = p_{y0} \forall i$). The marginal PDP (MPDP), such that the NCPS remains properly damped, i.e., $\forall p_{y0} >$MPDP, is given by $\sup\{\gamma > 0, \text{s.t. LMIs in (3.24) remain feasible}, \forall p_{y0} \in [\gamma, 1]\}$.

3.2.2 Physical significance of the developed LMIs

The physical meaning of the mathematical result given by the developed LMIs will be better understood using the concepts of observability and controllability. As mentioned in Section 3.1.1, the output measurements have high observability of the unstable and/or poorly damped electromechanical modes of the power system. The LQG controller requires the knowledge of these measurements and the state matrices to correctly estimate the states, which are then multiplied by the LQR gain to get the control input for the power system. The LQG controller requires the knowledge of these measurements and the state matrices in order to correctly estimate the states, which are then multiplied with the LQR gain to get the control input for the power system. The closed-loop system is properly stabilized and damped, provided the packet delivery rate is 100%. The decrease in packet delivery rate from 100% results in the loss of the output measurements in the communication network. The measurements which finally arrive at the controller after packet loss have an overall decrease in their observability for a given period of time, and the controller estimates the states with decreased accuracy. For a packet delivery rate of 0%, none of the measurements arrive at the LQG controller, and thus the observability is zero and the controller cannot estimate the states at all.

This concept of probabilistic observability will be better understood with an example. For example, if there are two measurements which are sent over the network, then there are four possibilities in a given time sample: (1) none of the measurements arrive, (2) only the first measurement arrives, (3) only the second measurement arrives, and (4) both measurements arrive. Each of these four possibilities has a probability associated with it depending on the packet delivery rates of the two communication channels. The overall observability of the arriving measurements depends on these four probabilities, and is thus a probabilistic quantity in itself. For q measurements, there are 2^q possibilities, and the overall observability will depend on all of these possibilities. A similar analogy applies for the controllability of the power system by the control inputs sent over the communication network, and thus the overall controllability is also a probabilistic quantity. The stability and the $\mathcal{D}$-stability of the closed-loop system depend on these probabilistic observability and controllability and are written in mathematical forms as (3.23) and (3.24), respectively.

3.3 CASE STUDY: 68-BUS 16-MACHINE 5-AREA NCPS

3.3.1 System description

The 16-machine, 68-bus model test system, with parameters as given in Appendix A, has been used for the case study and is shown in Fig. 3.5.

This is a reduced-order equivalent of the interconnected New England test system (NETS) and New York power system (NYPS) of 1970s. NETS and NYPS are represented by a group of generators, while the power import from each of the three other neighboring areas are approximated by equivalent generator models (G14 to G16). NYPS needs to import around 1.5 GW from Area 5, for which a TCSC is installed on the 18–50 tie line. Percentage compensation of the TCSC needs to be dynamically controlled to control the reactance of the tie line. A detailed system description is available in Appendix A, which is used to simulate the NCPS model in MATLAB SIMULINK, with the following modifications: G13 is taken as the reference machine instead of G16; the gain for PSS on G9 is taken as

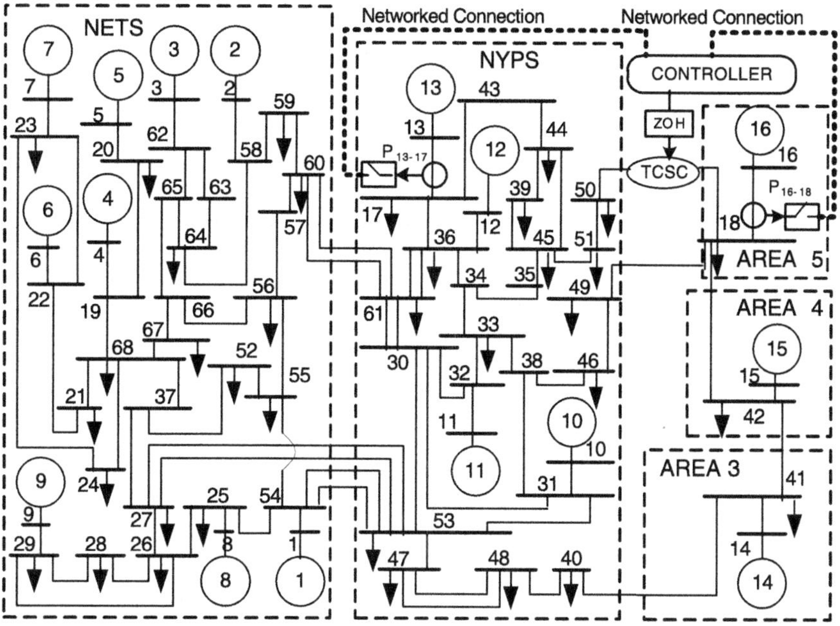

Figure 3.5 Line diagram of the 16-machine, 68-bus, 5-area NCPS.

20.5 p.u., with lead and lag time constants as 0.08 s and 0.03 s, respectively (for both stages); and the input signal given to the PSS is the p.u. rotor slip speed of G9 with respect to the reference machine (note that for rest of the chapters, the input signal given to the PSS is taken as p.u. rotor slip speed with respect to the synchronous speed of 1 p.u.).

3.3.2 Simulation results and discussion

3.3.2.1 Operating condition 1 (base case)

For the first case of system operation (total tie line flow between NETS and NYPS = 700 MW, no line outages), the damping and the frequency of the three poorly damped modes of the linearized system have been computed. The normalized participation factors (PFs) of all the states in these modes have also been calculated and arranged in decreasing order, as explained in Chapter 2. Table 3.1 gives the normalized PF of the top four states in these modes. As one can see in Table 3.1, the three poorly damped modes are indeed the interarea modes as they have strong participation from the electrodynamical states (rotor angle and speed) of all the three generators

Table 3.1 Normalized participation factors of the top four states in the three modes

Mode 1, $\zeta = 0.020$, $f = 0.394$ Hz		Mode 2, $\zeta = 0.041$, $f = 0.505$ Hz		Mode 3, $\zeta = 0.032$, $f = 0.598$ Hz	
State	**PF**	**State**	**PF**	**State**	**PF**
δ_{16}	1.000	Slip_{15}	1.000	Slip_{14}	1.000
Slip_{16}	0.999	δ_{15}	0.999	δ_{14}	0.999
δ_{15}	0.936	Slip_{14}	0.727	Slip_{6}	0.493
Slip_{15}	0.935	δ_{14}	0.726	δ_{6}	0.492

Table 3.2 Normalized residues of the active power flows in the three modes

Mode 1, $\zeta = 0.020$, $f = 0.394$ Hz		Mode 2, $\zeta = 0.041$, $f = 0.505$ Hz		Mode 3, $\zeta = 0.032$, $f = 0.598$ Hz	
Signal	**Residue**	**Signal**	**Residue**	**Signal**	**Residue**
P_{13-17}	1.000	P_{16-18}	1.000	P_{13-17}	1.000
P_{51-45}	0.773	P_{14-41}	0.760	P_{17-36}	0.698
P_{51-50}	0.665	P_{42-18}	0.727	P_{43-17}	0.600

G14, G15, and G16, which model the power generation in three different areas.

The open-loop system response confirms that the other electromechanical modes including one interarea mode of the system settles in less than 10 seconds and hence they have been left from the consideration of providing additional damping. The remote feedback signals are chosen based on residue analysis [88] for various active line power signals, as explained in Chapter 2. Table 3.2 gives the normalized residues of the top three active power flows in the three interarea modes. There are other means of robust signal selection to obtain the best signal(s) out of all the available signals, as described in [89], [90], and [91], to guarantee effectiveness of the signals for various operating scenarios.

The signals P_{13-17} (having the highest residues for modes 1 and 3) and P_{16-18} (having the highest residue for mode 2) have been selected as output signals (here P_{13-17} denotes the active power flow in the line from bus number 13 to bus number 17). With these two signals as output and Δk_{c-ss} (the control signal of the TCSC) as the input, the open-loop system is linearized to find the state space matrices. The system order is reduced (Section 3.1.1) to the lowest possible order such that the reduced system still remains a very good approximation of the full system in the frequency

Table 3.3 Comparison of modes for the full vs. the reduced system

	Frequency (in Hz)		Damping ratio	
Mode	**Full system**	**Reduced system**	**Full system**	**Reduced system**
1	0.394	0.394	0.020	0.020
2	0.505	0.500	0.041	0.046
3	0.598	0.598	0.032	0.034

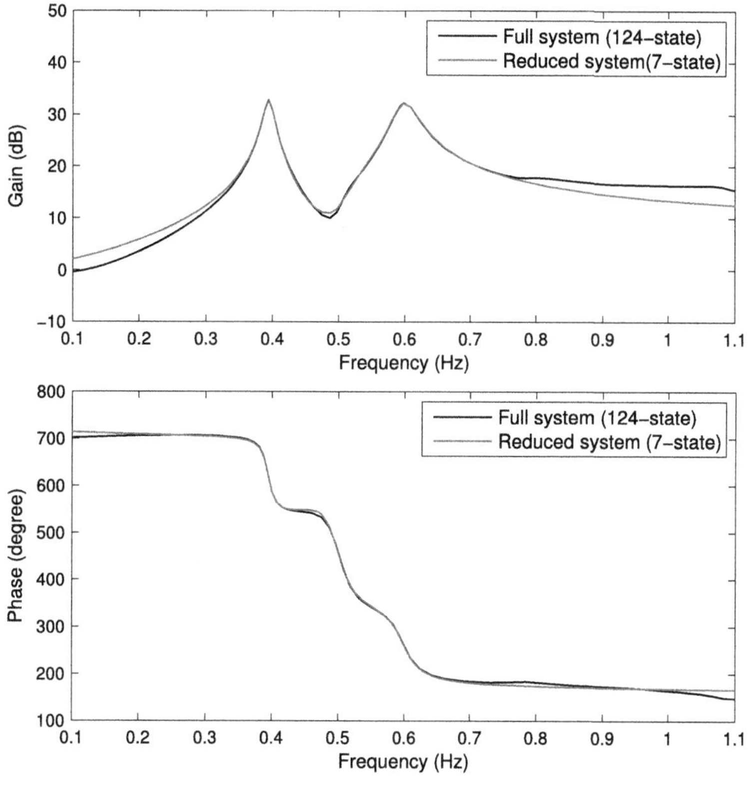

Figure 3.6 Frequency response of the full vs. the reduced system.

range of 0.2–1.0 Hz, and thus a reduced seventh-order system is obtained. Table 3.3, which compares frequencies and damping ratios of the three modes for the full and the reduced systems, and Fig. 3.6, which shows the frequency response of both the full and the reduced systems, prove that the reduced system is a good approximation of the full system in the frequency range of 0.2–1.0 Hz.

Remark. Deciding the sampling period: The controller needs to observe the system in the range of 0.2–1.0 Hz; hence the minimum required sampling frequency is 2.0 Hz (i.e., a maximum allowed sampling period of 0.5 s) according to the Nyquist–Shannon sampling theorem. This upper limit of the sampling period is also the threshold requirement for average time delay of the packets; if the average delay is more than this upper limit, then the packet loss rate will be very high and the network will not support the communication needs of the system. In the case study, a conservative sampling period of $T_0 = 0.1$ s has been assumed.

The packet loss in the path of the input and output signals has been modeled as a Bernoulli process. So, $\boldsymbol{\alpha}_k = \alpha_k^1$, while $\boldsymbol{\beta}_{k+1} = diag(\beta_{k+1}^1, \beta_{k+1}^2)$, in (3.7). The steady-state controller gain for the reduced system is found using the results of Section 3.1.4 and the modified Kalman filter is modeled using the principle described in Section 3.1.5. In the simulation, after one second a disturbance is created in the NCPS model by a three-phase fault and immediate outage of one of the tie lines between buses 53–54.

The open-loop system is a minimum phase system (which means that all of its zeros and poles are in the left half plane); thus it is required to check only the damping response of the system for various packet drop rates. For $\boldsymbol{\alpha}_k$, $\boldsymbol{S}_p = \{\emptyset, \{1\}\}$, and $\boldsymbol{T}_p(\boldsymbol{r})$ has two modes, $\boldsymbol{T}_1(\emptyset)$ and $\boldsymbol{T}_1(\{1\})$. For $\boldsymbol{\beta}_k$, $q = 2$ and $\boldsymbol{S}_q = \{\emptyset, \{1\}, \{2\}, \{1,2\}\}$, and $\boldsymbol{T}_q(\boldsymbol{s})$ has four modes viz. $\boldsymbol{T}_2(\emptyset)$, $\boldsymbol{T}_2(\{1\})$, $\boldsymbol{T}_2(\{2\})$, and $\boldsymbol{T}_2(\{1,2\})$. The corresponding jump state matrices $\mathcal{A}(\boldsymbol{r}, \boldsymbol{s})$ are $\mathcal{A}(\emptyset, \emptyset)$, $\mathcal{A}(\emptyset, \{1\})$, $\mathcal{A}(\emptyset, \{2\})$, $\mathcal{A}(\emptyset, \{1,2\})$, $\mathcal{A}(\{1\}, \emptyset)$, $\mathcal{A}(\{1\}, \{1\})$, $\mathcal{A}(\{1\}, \{2\})$, and $\mathcal{A}(\{1\}, \{1,2\})$, and their probabilities of occurrences are $(1 - p_{u1})(1 - p_{y1})(1 - p_{y2})$, $(1 - p_{u1})p_{y1}(1 - p_{y2})$, $(1 - p_{u1})(1 - p_{y1})p_{y2}$, $(1 - p_{u1})p_{y1}p_{y2}$, $p_{u1}(1 - p_{y1})(1 - p_{y2})$, $p_{u1}p_{y1}(1 - p_{y2})$, $p_{u1}(1 - p_{y1})p_{y2}$, and $p_{u1}p_{y1}p_{y2}$, respectively. Using these parameters, eight pairs of LMIs in (3.24) have been obtained; Θ is taken as 84.3 degrees corresponding to the 10% damping line, as shown in Fig. 3.4. Assuming the same network characteristics for all the network channels, i.e., $p_{u1} = p_{y1} = p_{y2} = p_y$, the feasibility of the LMIs has been checked for various values of p_y using the LMI toolbox in MATLAB. The toolbox returned a minimum feasible value of $p_y = 0.81$, i.e., the LMIs were feasible for $0.81 < p_y < 1.0$.

As data loss is a random process, multiple simulations have been performed for a given value of marginal PDP. Fig. 3.7 shows the rotor slip response for G16 for 100 simulations at a marginal PDP of 0.81. The mean value of the rotor slip for the 100 simulations has also been plotted in Fig. 3.7. In the rest of the plots in the case study only the mean value of multiple simulations is shown.

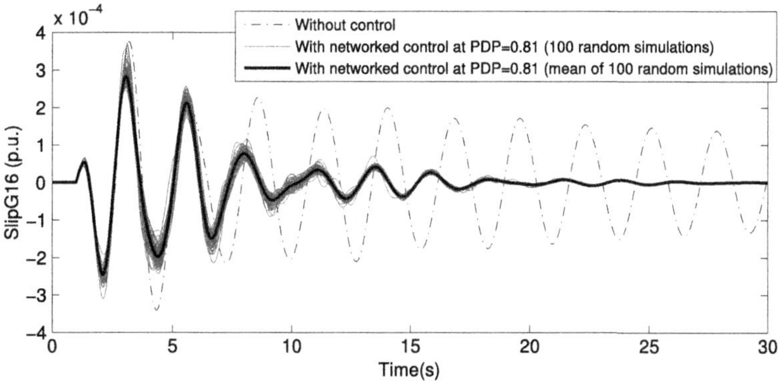

Figure 3.7 Rotor slip response for G16 at operating point 1.

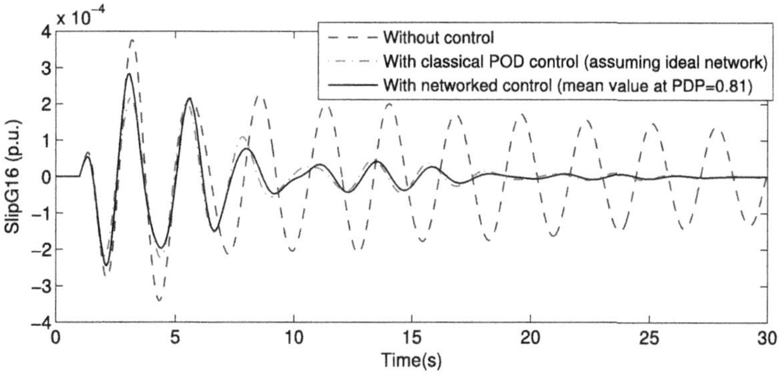

Figure 3.8 Comparison of rotor slip response at operating point 1.

System response using a classical damping controller (assuming a perfect communication link in its control loop, with infinite sampling rate and zero packet loss) has been shown for comparison in Fig. 3.8. Corresponding values of control signal are also shown in Fig. 3.9. The transfer function for the classical POD controller has been found using the theory discussed in Chapter 2 or in [65], and it is as follows (note that a washout filter has not been included in the designed POD controller, as a washout filter has not been used for the networked controller either):

$$k_{c\text{-}ss} = -\Delta P_{13\text{-}17} \times (\text{-}0.738) \left[\frac{1+0.138s}{1+0.725s}\right]^2 - \Delta P_{16\text{-}18} \times 0.925 \left[\frac{1+0.182s}{1+0.949s}\right]^2. \qquad (3.26)$$

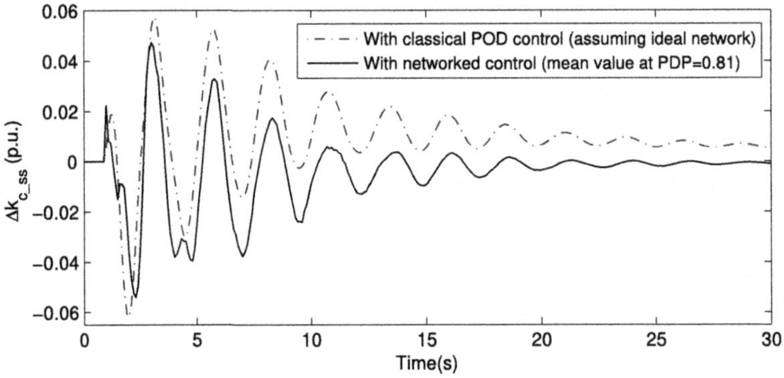

Figure 3.9 Comparison of control signals at operating point 1.

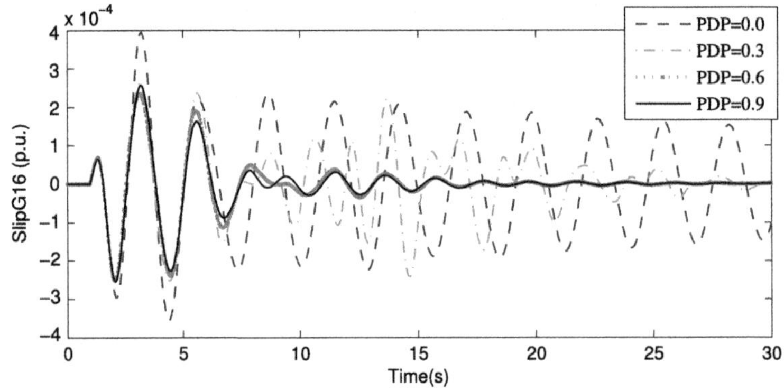

Figure 3.10 Rotor slip response for various packet delivery probabilities (PDPs).

The rotor slip response of G16 for the first operating point is also found for four other values of p_γ, as shown in Fig. 3.10.

Remark. It should be understood that it is not the sole purpose of Fig. 3.8 (and subsequent figures) to show that the performance of the networked controller is better than that of the classical damping controller. Rather, one another important purpose is to show that the performance of the networked control with communication packet dropout, even with marginal PDP, is comparable to the performance of classical control in which an ideal, lossless and delay-free communication network is assumed. Fig. 3.11 shows the comparison of the performance of networked control with that of classical control, when in both cases an ideal communication network is assumed (that is, PDP $= 1$). It can be clearly verified from the figure that the performance of networked control is much better than classical control

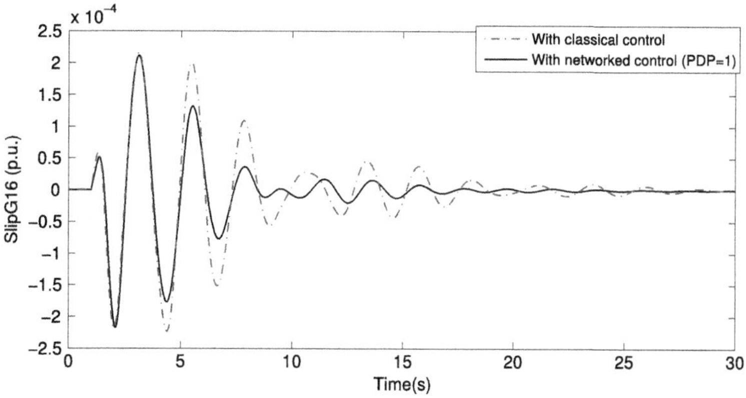

Figure 3.11 Classical vs. networked control, with assumption of an ideal network.

when ideal network conditions are assumed for both cases. Also, a metric which is used to assess the control effort required by a control method is the 2-norm of the output from the controller, or $\|\boldsymbol{u}\|_2$. The control effort for classical control is 0.32 p.u., while for networked control (with PDP $= 1$) it is 0.25 p.u. Thus networked control at 100% packet delivery rate is better than classical control, and it can damp the oscillations in a smaller amount of time, even when the control effort required by networked control is decreased by 22 % as compared to the control effort required by classical control.

3.3.2.2 Operating condition 2

In the second operating condition (total tie line flow between NETS and NYPS = 900 MW, no line outages), the open-loop system becomes unstable after the line outage, unlike the first operating condition. This is due to the presence of an unstable mode with negative damping ratio in the system. Therefore we can apply the LMI analysis (Section 3.2.1.1) to find the marginal PDP which can ensure closed-loop stability of the NCPS. It is found that the stability of the NCPS under this operating condition is ensured at a marginal PDP of 0.24, while the adequate damping of the system is ensured at a marginal PDP of 0.87 (using Section 3.2.1.2). The slip response of G16 is shown at both these marginal PDPs in Fig. 3.12.

It is evident in Fig. 3.8 and Fig. 3.10, for $T_0 = 0.1\ s$ and PDPs more than or equal to 0.81, the interarea modes of the system are properly damped. Similarly, it may be observed from Fig. 3.12 that the system in second oper-

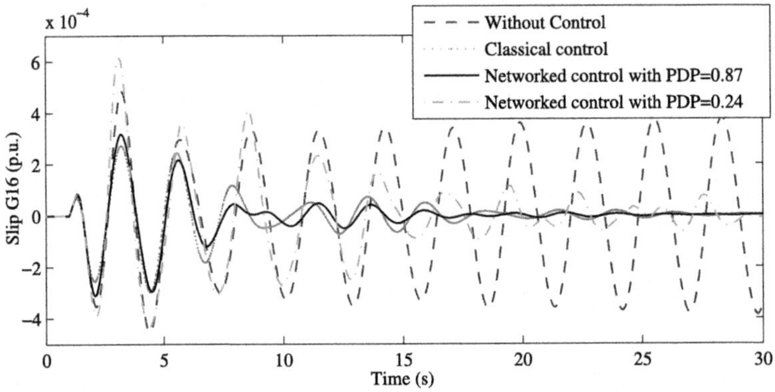

Figure 3.12 Rotor slip response for G16 at operating point 2.

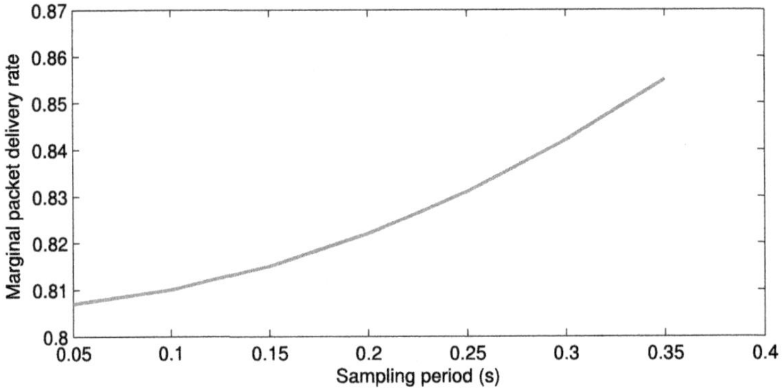

Figure 3.13 Marginal delivery probability vs. sampling period.

ating condition is stabilized at the marginal PDP of 0.24 while adequately damped at the marginal PDP of 0.87. So the results of the LMI analysis stand verified. It is also clear from Fig. 3.8 and Fig. 3.12 that the performance of the networked controller is better than the classical damping controller, at realistic packet delivery qualities that can be easily delivered by present-day telecom networks.

3.3.2.3 Effect of sampling period

Next, the effect of the sampling period on the marginal delivery probability for $\mathcal{D}$-stability is investigated. Fig. 3.13 shows the plot of the marginal delivery probability vs. the sampling period.

Table 3.4 Marginal packet delivery probability vs. operating point

S. No.	Total tie line flow (MW)	Line outage	Marginal PDP
1.	700	no outage	0.81
2.	700	60–61	0.83
3.	700	27–53	0.81
4.	100	no outage	0.79
5.	900	no outage	0.87
6.	100	27–53	0.79

One can easily infer from Fig. 3.13 that a longer sampling period requires an increase in p_y to guarantee feasibility. This is in line with the expectation that a packet has to be delivered with higher probability with an increase in sampling time.

3.3.2.4 Robustness

The robustness of the NCPS has been checked by obtaining the probabilities of marginal packet delivery for various operating conditions as listed in Table 3.4.

In Table 3.4, serial number 1 (S.No.1) is considered the base case of operation. For each operating condition, the control scheme is updated to give a corresponding LQG gain and reduced-order state space matrices for the Kalman filter. The stability performance of the NCPS has also been studied with a constant control scheme, i.e., the control scheme obtained for the base case is used for all the operating conditions. Fig. 3.14 shows the rotor slip response for three operating conditions with such constant control scheme.

It is clear from Table 3.4 and Fig. 3.14 that the NCPS is $\mathcal{D}$-stable for various operating points, even with a constant control scheme, at a feasible delivery probability of 0.85.

3.4 LIMITATIONS

The NCPS model developed in this chapter is a rigorous and generalized model, but it still suffers from various limitations and challenges.

The biggest implementational challenge for using the NCPS model for estimation and control of power system dynamics is posed by its strict communication network requirements. As has been mentioned earlier, the best of the currently available communication networks, the packet switching-based networks, even though fast (unlike the slow SCADA networks), suffer

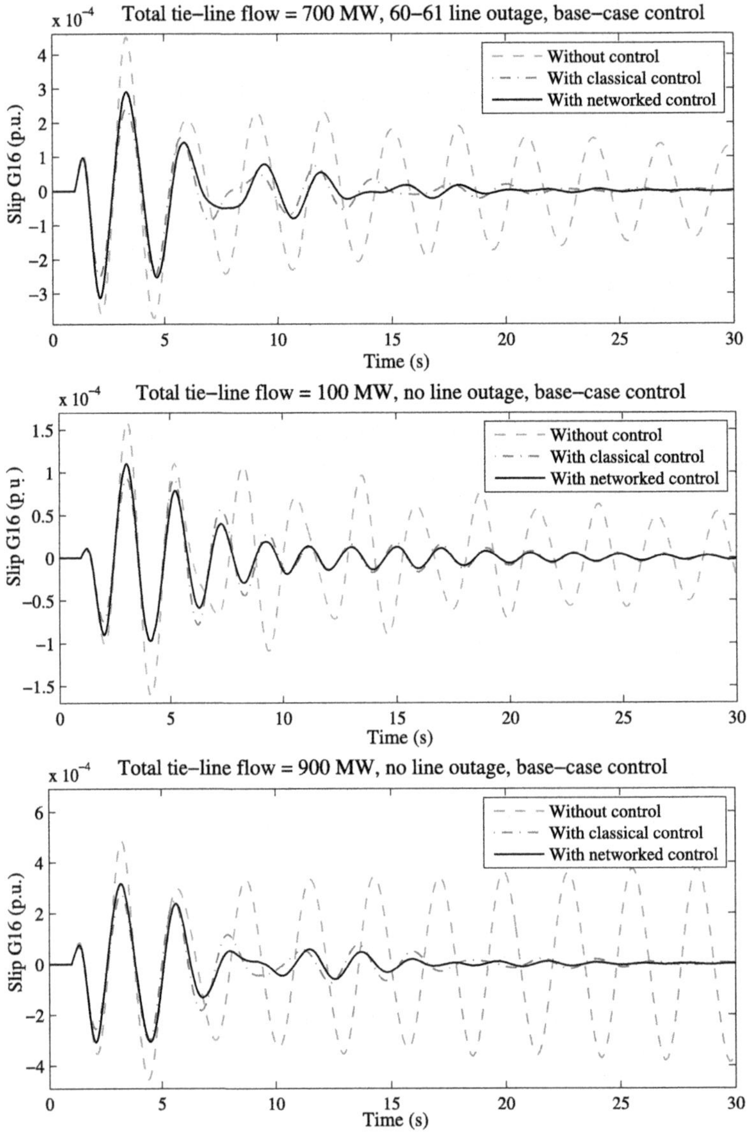

Figure 3.14 Rotor slip response for various operating points at $p_y = 0.85$.

from prohibitive drawbacks of random time delays, packet disordering, and cybersecurity threats, which can potentially destabilize the entire power system.

Some other limitations in the model presented are as follows.

1. The time delay model of the packets is integrated in the packet dropout model, as any packet with delay more than one sampling period is assumed to be dropped while the rest of the packets are deemed to be delivered successfully. A more accurate model for NCPS would have been the one in which the time delays and the packet dropouts are modeled independently.
2. The model is only valid as long as the operating conditions do not change significantly from the operating point at which the system is linearized. Hence, if, for instance, a generator goes out of service, then the transmission system operator (TSO) should get this information in real-time so that the system model and control gains are accordingly updated.
3. Linearization of the system at an operating point is nontrivial and requires an exact knowledge of steady-state values and system-wide parameters.

3.5 SUMMARY

This chapter has analyzed the stability effects of introducing a packet-based communication network in the control loops of a power system. A Kalman filtering and LQG-based scheme has been adopted for the centralized dynamic estimation and control framework for damping oscillatory modes of a power system. The random loss in the delivery of the packets has been modeled as a stochastic process. Using the developed LMIs, the lower limit on the probability of packet delivery has been computed which guarantees specified damping. It is found that under varying operating conditions the performance of the NCPS is robust.

The main idea of this chapter lies in the development of a generalized framework to assess the stability and damping of an NCPS. It also presents a formal approach for finding the minimum network requirements in terms of packet delivery quality, so that the specified stability and damping margins can be ensured for any operating condition of a power system. Specifically, the ideas may be summarized as follows.

1. A detailed characterization of the packet transmission process and the probability of packet loss have been considered in the framework of NCS for power system dynamic estimation and control in a centralized manner.
2. A practical output feedback methodology has been used for control (instead of state feedback), and the signals which are required to be

transmitted to the control unit are measurable line power signals. Also, a detailed and realistic subtransient power system model has been used.

3. The optimal control scheme which has been used for centralized control can be easily integrated with the WAMS or FACTS devices already present in the system.

The theory and application examples demonstrate that although the centralized dynamic estimation and control framework has limitations, it still has a good potential to guarantee a small signal stability margin for modern power systems.

CHAPTER 4

Decentralized Dynamic Estimation Using PMUs

As mentioned in Chapter 3, update rates of communication networks which are used in present-day power systems are not as fast as those assumed in the centralized dynamic state estimation (DSE) model. Moreover, even though a sufficiently fast communication network architecture based on packet switching can fulfill the requirement of high update rates, it suffers from other drawbacks (like packet dropout, random time delays, cybersecurity threats, etc.), which, if not rectified, render it unsuitable for use in power systems. This chapter, therefore, presents an alternative algorithm for DSE, one which eliminates the requirement of a communication network: the method of *decentralized* DSE (the method presented here uses phasor measurement units [PMUs]; a different method without the use of PMUs is presented in Chapter 6).

Many of the proposed methods for DSE are based on linear schemes (see, for example, [25] and [27]). These schemes involve linearization of a system's differential and algebraic equations (DAEs), followed by finding the Jacobian matrices. Linearization introduces approximation errors, which may become significant over time, especially for a complex and high-order power system model [30]. Moreover, calculation of Jacobian matrices is computationally expensive, as it has to be done at every iteration of the algorithm.

The drawbacks of linear schemes have been addressed in many research papers, which propose application of unscented transformation to eliminate linearization and calculation of Jacobians [28–31]. In [28], an unscented Kalman filter (UKF)-based algorithm has been proposed for DSE of a synchronous machine connected to an infinite bus (also called single-machine infinite-bus (SMIB) system). In [29], DSE of a SMIB system is performed using an extended particle filter. The SMIB system is an idealized approximation of a power system, which ignores much of the complexities of real power systems. This limitation has been addressed in [30], which proposes a centralized UKF algorithm for DSE of a multimachine power system. This algorithm requires that remote signals from all the machines in the system are transmitted to a central location. This method has its own limitations;

Dynamic Estimation and Control of Power Systems
https://doi.org/10.1016/B978-0-12-814005-5.00015-7

many of the signals required for estimation, such as rotor speed and state variables of the excitation system, are difficult to measure. Even if these signals are measured somehow, it is difficult to ensure their transmission to a central location at a high sampling rate. The decentralized method of DSE presented in this chapter addresses these issues.

The rest of the chapter is organized as follows. The problem statement and an overview of the DSE scheme are given in Section 4.1. Section 4.2 presents a description of discrete DAEs for power systems while the concept of decentralization is explained in Section 4.3. The theory of unscented Kalman filters and an algorithm for decentralized DSE are given in Section 4.4. Section 4.5 presents a case study of a 68-bus test model and Section 4.6 presents an algorithm for bad-data detection. Section 4.7 discusses other methods of decentralized DSE based on DSE, while Section 4.8 summarizes the chapter.

4.1 PROBLEM STATEMENT AND METHODOLOGY IN BRIEF

It is assumed that the power system is represented using a set of continuous-time nonlinear DAEs, given by

$$\begin{aligned} \dot{\boldsymbol{x}}(t) &= \bar{\boldsymbol{g}}[\boldsymbol{x}(t), \boldsymbol{u}(t), \boldsymbol{y}_a(t)] + \bar{\boldsymbol{v}}(t), \\ 0 &= \bar{\boldsymbol{h}}[\boldsymbol{x}(t), \boldsymbol{u}(t), \boldsymbol{y}_a(t)], \quad \boldsymbol{y}(t) = \boldsymbol{h}[\boldsymbol{x}(t), \boldsymbol{u}(t), \boldsymbol{y}_a(t)] + \boldsymbol{w}(t). \end{aligned} \tag{4.1}$$

After sampling (4.1) at a sampling period T_0, one gets

$$\begin{aligned} \frac{\boldsymbol{x}(kT_0) - \boldsymbol{x}((k-1)T_0)}{T_0} &= \bar{\boldsymbol{g}}[\boldsymbol{x}((k-1)T_0), \boldsymbol{u}((k-1)T_0), \boldsymbol{y}_a((k-1)T_0)] \\ &\quad + \bar{\boldsymbol{v}}((k-1)T_0), \end{aligned} \tag{4.2}$$

$$0 = \bar{\boldsymbol{h}}[\boldsymbol{x}(kT_0), \boldsymbol{u}(kT_0), \boldsymbol{y}_a(kT_0)], \tag{4.3}$$

$$\boldsymbol{y}(kT_0) = \boldsymbol{h}[\boldsymbol{x}(kT_0), \boldsymbol{u}(kT_0), \boldsymbol{y}_a(kT_0)] + \boldsymbol{w}(kT_0). \tag{4.4}$$

Rewriting kT_0 as k and $(k-1)T_0$ as $k-1$, (4.4) converts into a discrete form given by (4.6)–(4.8). We have

$$\boldsymbol{x}(k) = \boldsymbol{x}(k-1) + T_0\bar{\boldsymbol{g}}[\boldsymbol{x}(k-1), \boldsymbol{u}(k-1), \boldsymbol{y}_a(k-1)] + T_0\bar{\boldsymbol{v}}(k-1) \tag{4.5}$$

$$\Rightarrow \boldsymbol{x}(k) = \boldsymbol{g}[\boldsymbol{x}(k-1), \boldsymbol{u}(k-1), \boldsymbol{y}_a(k-1)] + \boldsymbol{v}(k-1), \tag{4.6}$$

$$0 = \bar{\boldsymbol{h}}[\boldsymbol{x}(k), \boldsymbol{u}(k), \boldsymbol{y}_a(k)], \tag{4.7}$$

$$\boldsymbol{y}(k) = \boldsymbol{h}[\boldsymbol{x}(k), \boldsymbol{u}(k), \boldsymbol{y}_a(k)] + \boldsymbol{w}(k). \tag{4.8}$$

In state estimation the state $\boldsymbol{x}(k)$ is treated as a random variable with an estimated mean $\hat{\boldsymbol{x}}(k)$ and an estimated covariance $\boldsymbol{P}_x(k)$.

The terms $\boldsymbol{v}(k)$ and $\boldsymbol{w}(k)$ are assumed to be white Gaussian noises. The constant covariance matrices for the noises are denoted as $\boldsymbol{P}_v$ for $\boldsymbol{v}(k)$ and $\boldsymbol{P}_w$ for $\boldsymbol{w}(k)$.

Remark. Although white Gaussian noises are used in this chapter, other types of noises may also be used (such as colored noises) as the UKF remains applicable in a wide variety of noise models, as shown in [92].

4.1.1 Problem statement

Find $\hat{\boldsymbol{X}}(k)$ and $\boldsymbol{P}_X(k)$, given $\hat{\boldsymbol{X}}(k-1)$, $\boldsymbol{P}_X(k-1)$, $\boldsymbol{g}$, $\bar{\boldsymbol{h}}$, $\boldsymbol{h}$, $\boldsymbol{u}(k-1)$, $\boldsymbol{u}(k)$, $\boldsymbol{y}_a(k-1)$, $\boldsymbol{y}_a(k)$, $\boldsymbol{y}(k)$, $\boldsymbol{P}_v$, and $\boldsymbol{P}_w$, under the constraints that:

- the algorithm is decentralized, that is, the algorithm for one generation unit should work independently from the algorithms for other units; and
- only those measurements may be used which are easily measurable using PMUs and are locally available.

Stating the problem in simpler terms, an iterative algorithm for finding real-time estimates of the mean and covariance of the states needs to be devised, provided the system DAEs, the inputs, the local PMU measurements, and all the noise covariances are available. The algorithm should be such that the estimation process for each generation unit remains independent of other units.

4.1.2 Methodology

A block diagram of the system and the decentralized methodology for finding a solution for the aforementioned problem statement is shown in Fig. 4.1.

Each generation unit is equipped with a PMU responsible for measuring various phasors associated with that unit, specifically voltage and current phasors. Power systems usually operate nearly at a constant system frequency of 50 or 60 Hz, and thus all the measured signals from the system have a fundamental harmonic component which is equal to the system frequency. Assuming that other harmonics are present in relatively small

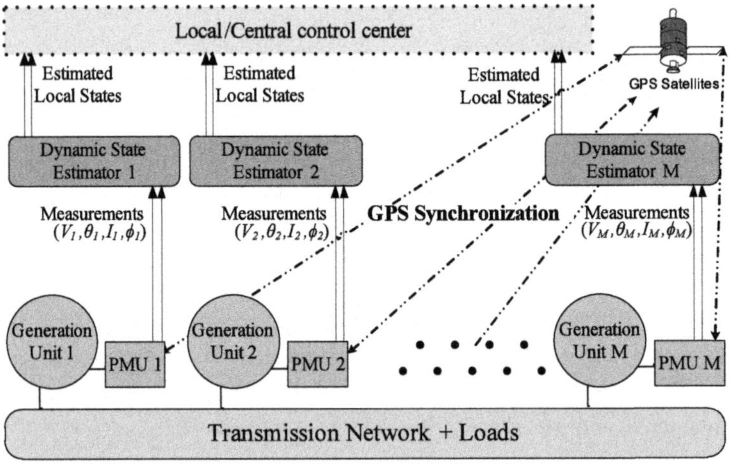

Figure 4.1 System block diagram and an overview of the methodology.

quantities, when the measured signals are sampled at more than twice the system frequency, the sampling does not lead to any loss of information in the signals, as per Nyquist–Shannon sampling theorem. PMUs provide sampling rates of over 600 Hz [17]; and hence they are capable of preserving signal information for state estimation purposes.

Any measuring device in a system (such as a PMU) has finite accuracy. This finite accuracy for a given measurement represented in the model as white Gaussian noise superimposed over a correct value of the signal. Each noise is assumed to have a zero mean and a standard deviation equal to the accuracy of the corresponding measurement. The sampled measurements, along with their noise variances, are sent from the PMU to a local estimator. The estimator is located in the vicinity of the PMU and hence communication requirements are assumed to be easily met. DSE is performed at the estimator using nonlinear unscented transformation in association with Kalman-like filtering. Estimates of all the dynamic states of the machine are then sent to local and/or central control centers for taking control decisions.

4.2 POWER SYSTEM MODELING AND DISCRETE DAES

Discrete DAEs of the power system are derived using continuous-time DAEs given in Chapter 2, and a brief description of the various components of the system is provided here.

4.2.1 Generators

As also explained in Chapter 2, each generator in the system has been represented using the subtransient model [62]. Slow speed governor dynamics have been ignored as they have practically no influence on the fast small-signal oscillatory dynamics of a power system [2]. Thus, mechanical torque, T_m, has been treated as a constant parameter. If T_m and other parameters for the machine (such as H and D) are not known, they may be estimated in real-time using the parameter estimation algorithm given in [92] or [93]. Discrete DAEs for the ith generator are given by (4.9)–(4.20). We have

$$\delta_i(k+1) = \delta_i(k) + T_0\omega_b(\omega_i(k) - 1), \tag{4.9}$$

$$\omega_i(k+1) = \omega_i(k) + \frac{T_0}{2H_i}\left(T_{mi} - T_{ei}(k) - D_i\left(\omega_i(k) - 1\right)\right), \tag{4.10}$$

$$\begin{aligned} E'_{qi}(k+1) = E'_{qi}(k) + \frac{T_0}{T'_{d0i}}[E_{fdi}(k) - E'_{qi}(k) \\ + (X_{di} - X'_{di})\{K_{d1i}I_{di}(k) + K_{d2i}\frac{\Psi_{1di}(k) - E'_{qi}(k)}{X'_{di} - X_{li}}\}], \end{aligned} \tag{4.11}$$

$$\begin{aligned} E'_{di}(k+1) = E'_{di}(k) - \frac{T_0}{T'_{q0i}}[E'_{di}(k) \\ + (X_{qi} - X'_{qi})\{K_{q1i}I_{qi}(k) + K_{q2i}\frac{\Psi_{2qi}(k) + E'_{di}(k)}{X'_{qi} - X_{li}}\}], \end{aligned} \tag{4.12}$$

$$\Psi_{2qi}(k+1) = \Psi_{2qi}(k) + \frac{T_0}{T''_{q0i}}[I_{qi}(k)(X'_{qi} - X_{li}) - E'_{di}(k) - \Psi_{2qi}(k)], \tag{4.13}$$

$$\Psi_{1di}(k+1) = \Psi_{1di}(k) + \frac{T_0}{T''_{d0i}}[I_{di}(k)(X'_{di} - X_{li}) + E'_{qi}(k) - \Psi_{1di}(k)], \tag{4.14}$$

$$E'_{dci}(k+1) = E'_{dci}(k) + \frac{T_0}{T_{ci}}((X''_{di} - X''_{qi})I_{qi}(k) - E'_{dci}(k)), \tag{4.15}$$

where $T_{ei}(k)$, $I_{di}(k)$, and $I_{qi}(k)$ are algebraic functions of $E'_{di}(k)$, $E'_{qi}(k)$, $\Psi_{1di}(k)$, $\Psi_{2qi}(k)$, $E'_{dci}(k)$, $V_i(k)$, $\theta_i(k)$, and $\delta_i(k)$ and are given by

$$T_{ei}(k) = K_{q1i}E'_{di}(k)I_{di}(k) + K_{d1i}E'_{qi}(k)I_{qi}(k) + K_{d2i}\Psi_{1di}(k)I_{qi}(k)$$
$$- K_{q2i}\Psi_{2qi}(k)I_{di}(k) + (X''_{di} - X''_{qi})I_{di}(k)I_{qi}(k), \quad (4.16)$$

$$I_{di}(k) = [R_{ai}\{E'_{di}(k)K_{q1i} - \Psi_{2qi}(k)K_{q2i} + E'_{dci}(k) - V_{di}(k)\}$$
$$- X''_{di}\{E'_{qi}(k)K_{d1i} + \Psi_{1di}(k)K_{d2i} - V_{qi}(k)\}]/Z^2_{ai}, \quad (4.17)$$

$$I_{qi}(k) = [R_{ai}\{E'_{qi}(k)K_{d1i} + \Psi_{1di}(k)K_{d2i} - V_{qi}(k)\}$$
$$+ X''_{di}\{E'_{di}(k)K_{q1i} - \Psi_{2qi}(k)K_{q2i} + E'_{dci}(k) - V_{di}(k)\}]/Z^2_{ai}, \quad (4.18)$$

$$\text{where, } V_{di}(k) = -V_i(k)\sin(\delta_i(k) - \theta_i(k)), \quad (4.19)$$

$$V_{qi}(k) = V_i(k)\cos(\delta_i(k) - \theta_i(k)), \ i = 1, 2, \ldots, M. \quad (4.20)$$

4.2.2 Excitation systems

Each generation unit may be excited manually or by using an excitation system. Two types of excitation systems have been considered in the case study. Discrete DAEs for the IEEE-DC1A type of excitation systems are given by (4.21)–(4.23), while for the IEEE-ST1A type of excitation systems they are given by (4.24)–(4.25). In the case of manual excitation, the field excitation voltage, E_{fd}, is equal to a constant reference, V_{ref}; we have

$$V_{ri}(k+1) = V_{ri}(k) + \frac{T_0}{T_{ri}}[V_i(k) - V_{ri}(k)], \quad (4.21)$$

$$V_{ai}(k+1) = V_{ai}(k) + \frac{T_0}{T_{ai}}[K_{ai}(V_{refi} + V_{ssi}(k) - V_{ri}(k)) - V_{ai}(k)], \quad (4.22)$$

$$E_{fdi}(k+1) = E_{fdi}(k) - \frac{T_0}{T_{xi}}[E_{fdi}(k)(K_{xi} + A_{xi}e^{B_{xi}E_{fdi}(k)}) - V_{ai}(k)],$$
$$E_{fdmini} \le E_{fdi}(k+1) \le E_{fdmaxi}, \ i = 1, 2, \ldots, M, \quad (4.23)$$

$$V_{ri}(k+1) = V_{ri}(k) + \frac{T_0}{T_{ri}}[V_i(k) - V_{ri}(k)], \quad (4.24)$$

$$E_{fdi}(k+1) = [K_{ai}(V_{refi} + V_{ssi}(k+1) - V_{ri}(k+1))],$$
$$E_{fdmini} \le E_{fdi}(k+1) \le E_{fdmaxi}, \ i = 1, 2, \ldots, M. \quad (4.25)$$

4.2.3 Power system stabilizer (PSS)

A discrete form of the transfer function of a power system stabilizer (PSS) (stated in Chapter 2) is given by (4.26)–(4.32). We write

$$P_{s1i}(k+1) = P_{s1i}(k) + \frac{T_0}{T_{wi}} P'_{s1i}(k), \tag{4.26}$$

$$P_{s2i}(k+1) = P_{s2i}(k) + \frac{T_0}{T_{12i}} P'_{s2i}(k), \tag{4.27}$$

$$P_{s3i}(k+1) = P_{s3i}(k) + \frac{T_0}{T_{22i}} P'_{s3i}(k), \tag{4.28}$$

$$\text{where, } P'_{s1i}(k) = K_{pssi}(\omega_i(k) - 1) - P_{s1i}(k), \tag{4.29}$$

$$P'_{s2i}(k) = P'_{s1i}(k) - P_{s2i}(k), \tag{4.30}$$

$$P'_{s3i}(k) = P'_{s1i}(k) + \frac{T_{11i} - T_{12i}}{T_{12i}} P'_{s2i}(k) - P_{s3i}(k), \tag{4.31}$$

$$\text{and } V_{ssi}(k) = P'_{s1i}(k) + \frac{T_{11i} - T_{12i}}{T_{12i}} P'_{s2i}(k) + \frac{T_{21i} - T_{22i}}{T_{22i}} P'_{s3i}(k),$$
$$V_{ssmini} \leq V_{ssi}(k+1) \leq V_{ssmaxi}, \quad i = 1, 2, \ldots, M. \tag{4.32}$$

4.2.4 Network model

Network current balance equations for the generator buses are given in discrete form by (4.33), while power balance equations for the nongenerator buses are given in discrete form by (4.34). We have

$$(I_{qi}(k) + jI_{di}(k))e^{j\delta_i(k)} = I_i(k)e^{j\phi_i(k)} = \mathbf{Y}_i(k)\boldsymbol{V}(k) + \frac{P_{Li}(k) - jQ_{Li}(k)}{V_i(k)e^{-j\theta_i(k)}},$$
$$\text{where } \mathbf{Y}_i(k) \text{ is the } i\text{th row of } \mathbf{Y}(k) \text{ and } i = 1, 2, \ldots, M, \tag{4.33}$$

$$(P_{Li}(k) - jQ_{Li}(k)) + V_i(k)e^{-j\theta_i(k)}(\mathbf{Y}_i(k)\boldsymbol{V}(k)) = 0,$$
$$\text{where } \mathbf{Y}_i(k) \text{ is the } i\text{th row of } \mathbf{Y}(k) \text{ and } i = (M+1), (M+2), \ldots, N. \tag{4.34}$$

4.3 PSEUDOINPUTS AND DECENTRALIZATION OF DAES

A generation unit consists of a generator, its excitation system, and a PSS when present. The DAEs for a unit, given by (4.9)–(4.32), are coupled to the DAEs for other units through the network equations given by (4.33)–(4.34). Inputs to the power system come from system disturbances, such as load changes, line faults, and generation failures. If it is assumed that none of the dynamic states are directly measured, a centralized state estimation scheme would require real-time information about all system-wide disturbances, besides information of line parameters, parameters for all the generation units, and system-wide PMU measurements. Obtaining such real-time information is practically not feasible. A decentralized scheme of estimation is the only practical alternative.

An inspection of (4.9)–(4.32) reveals that the ith generation units I, ϕ, and the dynamic states for the $(k+1)$th sample are explicit functions of V, θ, and the dynamic states for the kth sample. This inspection leads to an idea which forms the basis of the decentralized estimation scheme: *if V and θ are treated as inputs, rather than as measurements, and I and ϕ are treated as outputs (that is, as normal measurements), then the dynamic equations for one generation unit can be decoupled from the dynamic equations for other units.* This idea of "pseudoinputs" forms the central theme of the concept of decentralization. It must be noted here that this representation is not unique, and the DAEs can be rearranged in such a way that V and θ become the outputs and I and ϕ become the inputs. The idea is, therefore, to use one of the pairs of measurements as an input pair and the other pair as an output pair. In this chapter V and θ are treated as the input pair.

The idea of pseudoinputs may be better understood with a simpler model of a power system. A classical model of a power system in discrete form is given by the following DAEs for the ith machine:

$$\delta_i(k+1) = \delta_i(k) + T_0\omega_b(\omega_i(k) - 1), \tag{4.35}$$

$$\omega_i(k+1) = \omega_i(k) + \frac{T_0}{2H_i}\left(T_{mi} - T_{ei}(k) - D_i\left(\omega_i(k) - 1\right)\right), \tag{4.36}$$

$$\text{where } T_{ei}(k) = E'_{qi}I_i(k)\cos(\delta_i(k) - \phi_i(k)) = \frac{E'_{qi}}{x'_d}V_i(k)\sin(\delta_i(k) - \theta_i(k)) \tag{4.37}$$

and (V, θ) and (I, ϕ) are related as $I_i(k)e^{j\phi_i(k)} = \dfrac{E'_{qi}e^{j\delta_i(k)} - V_i(k)e^{j\theta_i(k)}}{jx'_d}$.

$$(4.38)$$

Here E'_{qi} is treated as a constant parameter in the classical model. The various bus voltages and currents in the system are coupled by the same network equations as in the subtransient model (that is, by (4.33)–(4.34)). In the centralized method of DSE (such as in [30]), the central estimator requires a complete system model and real-time knowledge of all the changes/disturbances occurring in the system. When a disturbance occurs, the estimator predicts the new states of all the machines (in the state prediction step) and the new voltages and currents of all the buses (in the measurement prediction step) by incorporating the disturbance in the complete system model. The predicted values are then corrected using the measured values of bus voltages and currents in the Kalman update step, and thus new state estimates are generated.

In the decentralized method of DSE, each machine has its own estimator. Each decentralized estimator treats one of the pairs of (V, θ) and (I, ϕ) as input and the other pair as normal measurement, and hence requires only equations (4.35)–(4.37) for state prediction and (4.38) for measurement prediction. If the pair (V, θ) is used as pseudoinput, then $T_{ei}(k) = (E'_{qi}/x'_d)V_i(k)\sin(\delta_i(k) - \theta_i(k))$ is used in the state prediction step and $I_i(k)e^{j\phi_i(k)} = (E'_{qi}e^{j\delta_i(k)} - V_i(k)e^{j\theta_i(k)})/(jx'_d)$ is used in the measurement prediction step. Otherwise, if the pair (I, ϕ) is used as pseudoinput, then $T_{ei}(k) = E'_{qi}I_i(k)\cos(\delta_i(k) - \phi_i(k))$ is used in the state prediction step and $V_i(k)e^{j\theta_i(k)} = E'_{qi}e^{j\delta_i(k)} - jx'_dI_i(k)e^{j\phi_i(k)}$ is used in the measurement prediction step. Thus the network equations (4.33)–(4.34) are not required, and the machine equations are decoupled.

Physical significance of the above idea of decentralization may be understood by going deeper into the physics of power system dynamics. Any change or disturbance which takes place at one point in a large-scale power system is propagated quickly throughout the system. This is because the propagation of disturbances takes place over an electromechanical traveling wave which travels at a high speed and takes less than a second to propagate changes throughout all the bus voltages and currents in the system (as elaborated in the chapter on "Electromechanical Wave Propagation" in [16]). These changes in voltage and current levels are in fact responsible for initiating slower small-signal oscillatory dynamics of devices which are connected to the buses. Therefore, just the knowledge of local bus voltage and current is sufficient to predict and estimate the dynamics of the devices that are connected to that local bus; in our case this device is a synchronous generator. But this knowledge of local voltage and current must be complete (both magnitude and phase are required), and this makes the synchronization of various PMU devices through the GPS satellites crucial to the estimation process. This synchronization of PMUs may also be considered as an indirect coordination between the decentralized estimators.

The idea of decoupling by treating V and θ as inputs leads to a problem: only measured values of V and θ are available (given by V_y and θ_y, respectively), instead of their actual values, and hence they have associated noises, given by V_w and θ_w, respectively. One way of including these noises in the DAEs is to model them as input noises [94]. But this would require linearization and would therefore defeat the purpose of unscented transformation and nonlinear filtering. Another way of including the measurement noises is to redefine the values of V and θ according to (4.39), based on the fact that the actual inputs are equal to the differences of their measured values and the associated noises. We have

$$V_i(k) = V_{yi}(k) - V_{wi}(k); \;\; \theta_i(k) = \theta_{yi}(k) - \theta_{wi}(k). \tag{4.39}$$

If the expressions for $V_i(k)$ and $\theta_i(k)$ from (4.39) are used in (4.9)–(4.32), the resultant DAEs give the decentralized process model for the ith generation unit, which is written in the following form, with $\boldsymbol{x}_i$ as the vector of the dynamic states and $\boldsymbol{g}_i$ as the corresponding state functions:

$$\boldsymbol{x}_i(k) = \boldsymbol{g}_i[\boldsymbol{x}_i(k-1), \boldsymbol{u}'_i(k-1), \boldsymbol{z}_i(k-1)] + \boldsymbol{v}_i(k-1), \;\; i = 1, 2, \ldots, M. \tag{4.40}$$

In the above model, $\boldsymbol{u}'_i(k-1)$ acts as a *pseudoinput* vector, $\boldsymbol{z}_i(k-1)$ is its noise, and $\boldsymbol{u}'_i(k-1)$ and $\boldsymbol{z}_i(k-1)$ are given as

$$\boldsymbol{u}'_i(k-1) = [V_{yi}(k-1), \theta_{yi}(k-1)]^T, \tag{4.41}$$

$$\boldsymbol{z}_i(k-1) = [V_{wi}(k-1), \theta_{wi}(k-1)]^T, \;\; i = 1, 2, \ldots, M, \tag{4.42}$$

where V_{wi} and θ_{wi} are white noises with zero mean and constant standard deviations given by $\sigma_{V_{wi}}$ and $\sigma_{\theta_{wi}}$, respectively. Thus, the mean and covariance of $\boldsymbol{z}_i(k-1)$ also remain constant for each sample; for $i = 1, 2, \ldots, M$, they are given by

$$\hat{\boldsymbol{z}}_i(k-1) = \boldsymbol{0}_{2\times 1}, \boldsymbol{P}_{zi}(k-1) = \boldsymbol{P}_{zi} = \mathrm{diag}\{\sigma_{V_{wi},}{}^2 \sigma_{\theta_{wi}}{}^2\}. \tag{4.43}$$

If $\hat{\boldsymbol{x}}_i(k-1)$ and $\boldsymbol{P}_{xi}(k-1)$ are the estimates of mean and covariance of $\boldsymbol{x}_i(k-1)$, $\boldsymbol{P}_{xzi}(k-1)$ is the crosscorrelation between $\boldsymbol{x}_i(k-1)$ and $\boldsymbol{z}_i(k-1)$, and $\boldsymbol{x}_i(k-1)$ is augmented with $\boldsymbol{z}_i(k-1)$ to give $\mathbf{X}_i(k-1) = [\boldsymbol{x}_i(k-1)^T, \boldsymbol{z}_i(k-1)^T]^T$, then the estimates of mean and covariance of $\mathbf{X}_i(k-1)$, for $i = 1, 2, \ldots, M$, are given by

$$\hat{\mathbf{X}}_i(k-1) = \begin{bmatrix} \hat{\boldsymbol{x}}_i(k-1) \\ \hat{\boldsymbol{z}}_i(k-1) \end{bmatrix}, \tag{4.44}$$

$$\boldsymbol{P}_{Xi}(k-1) = \begin{bmatrix} \boldsymbol{P}_{xi}(k-1) & \boldsymbol{P}_{xzi}(k-1)^T \\ \boldsymbol{P}_{xzi}(k-1) & \boldsymbol{P}_{zi}(k-1) \end{bmatrix}. \tag{4.45}$$

The augmented state $\boldsymbol{X}(k)$ is also a random variable with an estimated mean $\hat{\boldsymbol{X}}(k)$ and an estimated covariance $\boldsymbol{P}_X(k)$. Rewriting (4.40) in the augmented state form, one gets

$$\mathbf{X}_i(k) = \boldsymbol{g}_i[\mathbf{X}_i(k-1), \boldsymbol{u}'_i(k-1)] + \begin{bmatrix} \boldsymbol{v}_i(k-1) \\ \mathbf{0}_{2\times 1} \end{bmatrix}, \quad i = 1, 2, \ldots, M. \tag{4.46}$$

Measurement equations for the measured magnitude, I_{yi}, and the measured phase, ϕ_{yi}, of the stator current of the ith unit are (using (4.33))

$$I_{yi}(k) = \sqrt{(I_{qi}(k))^2 + (I_{di}(k))^2} + I_{wi}(k), \tag{4.47}$$

$$\phi_{yi}(k) = \arg\{I_{qi}(k) + jI_{di}(k)\} + \delta_i(k) + \phi_{wi}(k), \quad i = 1, 2, \ldots, M. \tag{4.48}$$

In (4.47), $I_{qi}(k)$ and $I_{di}(k)$ are given by (4.20) after replacing the expressions of $V_i(k)$ and $\theta_i(k)$ from (4.39). Writing $[I_{yi}, \phi_{yi}]^T$ as the output vector $\boldsymbol{y}_i$, the corresponding measurement functions (given by (4.47), (4.39), and (4.20)) as $\boldsymbol{h}_i$, and $[I_{wi}, \phi_{wi}]^T$ as the output noise vector $\boldsymbol{w}_i$, the measurement model comes out as

$$\boldsymbol{y}_i(k) = \boldsymbol{h}_i[\mathbf{X}_i(k), \boldsymbol{u}'_i(k)] + \boldsymbol{w}_i(k), \quad i = 1, 2, \ldots, M. \tag{4.49}$$

The mean and covariance of $\boldsymbol{w}_i(k)$, for $i = 1, 2, \ldots, M$, are

$$\hat{\boldsymbol{w}}_i(k) = \mathbf{0}_{2\times 1}; \ \boldsymbol{P}_{wi}(k) = \boldsymbol{P}_{wi} = \mathrm{diag}\{\sigma_{I_{wi}}{}^2, \sigma_{\phi_{wi}}{}^2\}. \tag{4.50}$$

The aggregate model for the ith unit, given by (4.46) and (4.49), is the decentralized equivalent of (4.6)–(4.8).

4.4 UNSCENTED KALMAN FILTER (UKF)

Unscented transformation was proposed by J.K. Uhlmann as a general method for approximating nonlinear transformations of probability distributions [95,96]. Based on an idea that it is easier to approximate a probability distribution than to approximate a nonlinear function, this method is used to find consistent, efficient, and unbiased estimates of mean and covariance of a random variable undergoing a nonlinear transformation [97]. If the nonlinear transformation given by (4.46) is applied to $\mathbf{X}(k-1)$ (the suffix i has been ignored), then the estimated mean and covariance of the resultant state $\mathbf{X}(k)$ are derived in four steps, elaborated on in this section.

4.4.1 Generation of sigma points

The first step is to generate a set of points, called sigma points, whose sample mean and covariance are the same as that of $\mathbf{X}(k-1)$. If the dimension of $\mathbf{X}(k-1)$ is n, then just $2n$ sigma points, $\chi_l(k-1)$, $l=1,2,\ldots,2n$, need to be generated to capture its distribution [96]. The following algorithm is used for generation of the sigma points [28]:

$$\chi_l(k-1)=\hat{\mathbf{X}}(k-1)+\left(\sqrt{n\boldsymbol{P}_{\mathrm{X}}(k-1)}\right)_l,\ \ l=1,2,\ldots,n, \tag{4.51}$$

$$\chi_l(k-1)=\hat{\mathbf{X}}(k-1)-\left(\sqrt{n\boldsymbol{P}_{\mathrm{X}}(k-1)}\right)_l, l=(n+1),(n+2),\ldots,2n. \tag{4.52}$$

Here, $\left(\sqrt{n\boldsymbol{P}_{\mathrm{X}}(k-1)}\right)_l$ is the lth column of the lower triangular matrix $\sqrt{n\boldsymbol{P}_{\mathrm{X}}(k-1)}$ obtained by Cholesky decomposition, which is given by

$$n\boldsymbol{P}_{\mathrm{X}}(k-1)=\sqrt{n\boldsymbol{P}_{\mathrm{X}}(k-1)}\sqrt{n\boldsymbol{P}_{\mathrm{X}}(k-1)}^T. \tag{4.53}$$

4.4.2 State prediction

In the second step, predicted-state sigma points are generated, which are given by $\chi_l^-(k)=\boldsymbol{g}[\chi_l(k-1),\boldsymbol{u}'(k-1)]$, $l=1,2,\ldots,2n$. The sample mean of these points is equal to $\hat{\mathbf{X}}^-(k)$, while the sum of the augmented $\boldsymbol{P}_v$ and the sample covariance of these points is equal to $\boldsymbol{P}_{\mathrm{X}}^-(k)$. Here, $\hat{\mathbf{X}}^-(k)$ and $\boldsymbol{P}_{\mathrm{X}}^-(k)$ are estimated mean and estimated covariance, respectively, of a predicted-state random variable, $\mathbf{X}^-(k)$.

4.4.3 Measurement prediction

The third step is to generate predicted-measurement sigma points, which are given by $\boldsymbol{\gamma}_l^-(k) = \boldsymbol{h}[\boldsymbol{\chi}_l^-(k), \boldsymbol{u}'(k)]$, $l = 1, 2, \ldots, 2n$. The sample mean of these points is equal to $\hat{\boldsymbol{y}}^-(k)$, while the sum of $\boldsymbol{P}_w$ and the sample covariance of these points is equal to $\boldsymbol{P}_y^-(k)$. Here, $\hat{\boldsymbol{y}}^-(k)$ and $\boldsymbol{P}_y^-(k)$ are estimated mean and estimated covariance, respectively, of a predicted-measurement random variable, $\mathbf{y}^-(k)$. Crosscorrelation between the predicted-state sigma points and the predicted-measurement sigma points is equal to $\boldsymbol{P}_{Xy}^-(k)$, which is taken as estimated crosscorrelation between $\mathbf{X}^-(k)$ and $\mathbf{y}^-(k)$.

4.4.4 Kalman update

The final step is to find $\hat{\boldsymbol{X}}(k)$ and $\boldsymbol{P}_X(k)$ using the normal Kalman filter equations [98],

$$\boldsymbol{K}(k) = \boldsymbol{P}_{Xy}^-(k)(\boldsymbol{P}_y^-(k))^{-1}, \tag{4.54}$$

$$\hat{\boldsymbol{X}}(k) = \hat{\boldsymbol{X}}^-(k) + \boldsymbol{K}(k)(\boldsymbol{y}(k) - \hat{\boldsymbol{y}}^-(k)), \tag{4.55}$$

$$\boldsymbol{P}_X(k) = \boldsymbol{P}_X^-(k) - \boldsymbol{K}(k)[\boldsymbol{P}_{Xy}^-(k)]^T. \tag{4.56}$$

The above four steps constitute the UKF. As stated in the beginning of this chapter, the superiority of UKF has been established over other nonlinear filters, such as the extended Kalman filter [99].

Coming back to power systems, the aggregate model for one generation unit, given by (4.49) and (4.46), is completely independent from other units. Thus, the four steps of UKF may be directly applied to the ith aggregate model to give its filtering algorithm, see Algorithm 4.1 and Fig. 4.2.

4.5 CASE STUDY: 68-BUS TEST SYSTEM

The 16-machine, 68-bus test system, shown in Fig. 4.3, has been used for the case study. This system is similar to the one used in last chapter, the only difference being the absence of networked control and of any FACTS device in the system.

A detailed system description is available in Appendix A, which is used to simulate the system in MATLAB on a personal computer with Intel Core 2 Duo, 2.0 GHz CPU, and 2 GB RAM.

Algorithm 4.1 Decentralized DSE for the ith generation unit.

Begin Find $\boldsymbol{g}_i$, $\boldsymbol{h}_i$, $\boldsymbol{P}_{zi}$, and $\boldsymbol{P}_{wi}$ according to (4.46), (4.49), (4.43), and (4.50), respectively. Find $\boldsymbol{P}_{vi}$. Let m_i denote the total number of states to be estimated for the unit. Denote $n_i = m_i + 2$. Denote the steady-state values of $\hat{\boldsymbol{x}}_i$ as $\boldsymbol{x}_{0i}$.

While $(k \geq 1)$

{ **STEP 1: Initialize**

if $(k=1)$ **then** initialize $\hat{\boldsymbol{x}}_i(0) = \boldsymbol{x}_{0i}$, $\hat{\boldsymbol{z}}_i(0) = \boldsymbol{0}_{2\times 1}$, $\boldsymbol{P}_{xi}(0) = \boldsymbol{P}_{vi}$, $\boldsymbol{P}_{xzi}(0) = \boldsymbol{0}_{2\times m_i}$, $\boldsymbol{P}_{zi}(0) = \boldsymbol{P}_{zi}$ in (4.44) to get $\boldsymbol{P}_{Xi}(0)$ and $\hat{\boldsymbol{X}}_i(0)$.

else reinitialize $\hat{\boldsymbol{z}}_i(k-1) = \boldsymbol{0}_{2\times 1}$ and $\boldsymbol{P}_{zi}(k-1) = \boldsymbol{P}_{zi}$, leaving rest of the elements in $\hat{\boldsymbol{X}}_i(k-1)$ and $\boldsymbol{P}_{Xi}(k-1)$ unchanged.

STEP 2: Generate sigma points

$\boldsymbol{\chi}_{il}(k-1) = \hat{\boldsymbol{X}}_i(k-1) + (\sqrt{n_i \boldsymbol{P}_{Xi}(k-1)})_l, \ l = 1, 2, \ldots, n_i$

$\boldsymbol{\chi}_{il}(k-1) = \hat{\boldsymbol{X}}_i(k-1) - (\sqrt{n_i \boldsymbol{P}_{Xi}(k-1)})_l, l = (n_i+1), (n_i+2), \ldots, 2n_i$

STEP 3: Predict states

$\boldsymbol{\chi}_{il}^-(k) = \boldsymbol{g}_i[\boldsymbol{\chi}_{il}(k-1), \boldsymbol{u}_i'(k-1)], l = 1, \ldots, 2n_i; \ \hat{\boldsymbol{X}}_i^-(k) = \frac{1}{2n_i}\sum_{l=1}^{2n_i} \boldsymbol{\chi}_{il}^-(k)$

$\boldsymbol{P}_{Xi}^-(k) = \frac{1}{2n_i}\sum_{l=1}^{2n_i} [\boldsymbol{\chi}_{il}^-(k) - \hat{\boldsymbol{X}}_i^-(k)][\boldsymbol{\chi}_{il}^-(k) - \hat{\boldsymbol{X}}_i^-(k)]^T + \begin{bmatrix} \boldsymbol{P}_{vi} & \boldsymbol{0}_{m\times 2} \\ \boldsymbol{0}_{2\times m} & \boldsymbol{0}_{2\times 2} \end{bmatrix}$

STEP 4: Predict measurements

$\boldsymbol{y}_{il}^-(k) = \boldsymbol{h}_i[\boldsymbol{\chi}_{il}^-(k), \boldsymbol{u}_i'(k)], \ l = 1, 2, \ldots, 2n_i, \ \hat{\boldsymbol{y}}_i^-(k) = \frac{1}{2n_i}\sum_{l=1}^{2n_i} \boldsymbol{y}_{il}^-(k)$

$\boldsymbol{P}_{yi}^-(k) = \frac{1}{2n_i}\sum_{l=1}^{2n_i} [\boldsymbol{y}_{il}^-(k) - \hat{\boldsymbol{y}}_i^-(k)][\boldsymbol{y}_{il}^-(k) - \hat{\boldsymbol{y}}_i^-(k)]^T + \boldsymbol{P}_{wi}$

$\boldsymbol{P}_{Xyi}^-(k) = \frac{1}{2n_i}\sum_{l=1}^{2n_i} [\boldsymbol{\chi}_{il}^-(k) - \hat{\boldsymbol{X}}_i^-(k)][\boldsymbol{y}_{il}^-(k) - \hat{\boldsymbol{y}}_i^-(k)]^T$

STEP 5: Kalman update

$\boldsymbol{K}_i(k) = \boldsymbol{P}_{Xyi}^-(k)(\boldsymbol{P}_{yi}^-(k))^{-1}; \hat{\boldsymbol{X}}_i(k) = \hat{\boldsymbol{X}}_i^-(k) + \boldsymbol{K}_i(k)(\boldsymbol{y}_i(k) - \hat{\boldsymbol{y}}_i^-(k))$

$\boldsymbol{P}_{Xi}(k) = \boldsymbol{P}_{Xi}^-(k) - \boldsymbol{K}_i(k)[\boldsymbol{P}_{Xyi}^-(k)]^T$

STEP 6: Output and time update

output $\hat{\boldsymbol{X}}_i(k)$ and $\boldsymbol{P}_{Xi}(k)$

$k \leftarrow (k+1)$ }

There are three types of generation units in the test system. The first eight units in the system are of *type 1*: with the IEEE-DC1A type of excitation system and without a PSS. The ninth unit is of *type 2*: with the IEEE-ST1A type of excitation system and with a PSS installed. The rest of the units are of *type 3*: with manual excitation and without a PSS. The state vectors for the ith unit in the test system, according to these three types,

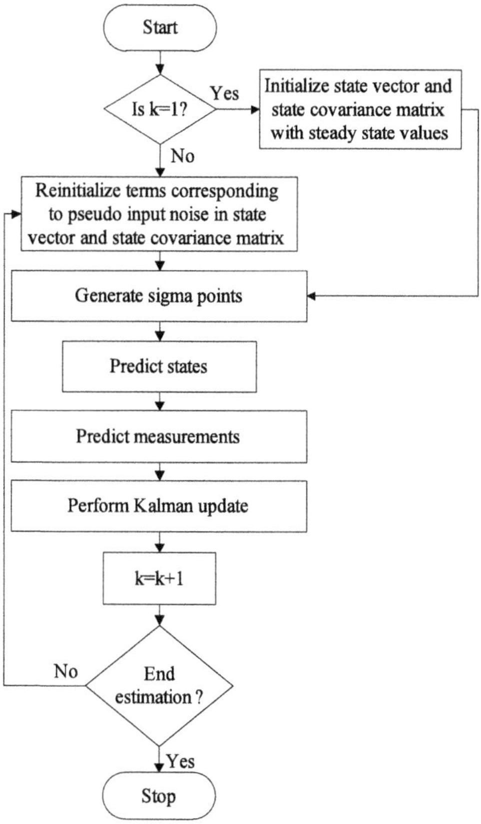

Figure 4.2 Flow chart for the steps of decentralized DSE.

are

$$\boldsymbol{x}_i = [\delta_i, \omega_i, E'_{qi}, E'_{di}, \psi_{2qi}, \psi_{1di}, V_{ri}, V_{ai}, E_{fdi}]^T, i = 1, 2, \ldots, 8, \tag{4.57}$$

$$\boldsymbol{x}_9 = [\delta_9, \omega_9, E'_{q9}, E'_{d9}, \psi_{2q9}, \psi_{1d9}, V_{r9}, P_{s1,9}, P_{s2,9}, P_{s3,9}]^T, \tag{4.58}$$

$$\boldsymbol{x}_i = [\delta_i, \omega_i, E'_{qi}, E'_{di}, \psi_{2qi}, \psi_{1di}]^T, i = 10, 11, \ldots, 16. \tag{4.59}$$

In the time-domain simulation, the actual values of V, θ, I, and ϕ were sampled at 120 Hz (T_0 = 8.33 ms), as the system frequency is taken to be 60 Hz for the 68-bus test system and 120 Hz is the Nyquist sampling frequency for this system frequency.

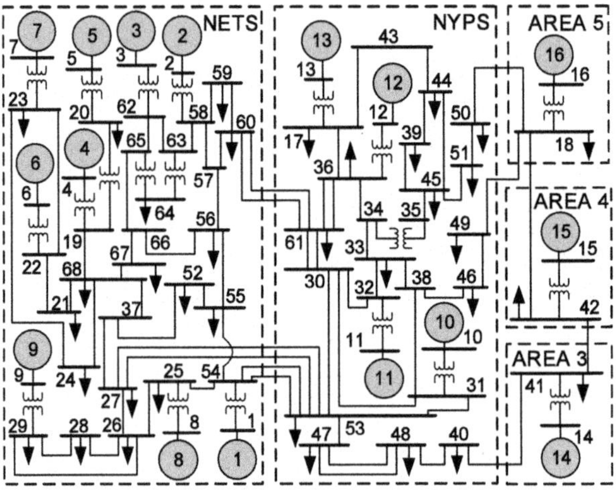

Figure 4.3 Line diagram of the 16-machine, 68-bus power system model.

4.5.1 Noise variances

4.5.1.1 Measurement noise

All of the PMUs in the power system are time-synchronized to an absolute time reference provided by GPS. The IEEE standard for synchrophasor measurements for power systems specifies a basic time synchronization accuracy of ±0.2 μs [18]. At 50 Hz, this translates to a phase measurement accuracy of around ±0.06 mrad. Thus, PMUs are expected to have an accuracy of around ±0.1 mrad for phase measurements. The accuracy of PMUs in magnitude measurements is limited by the accuracy of current transformers (CTs) and potential transformers (PTs) (also called instrument transformers). PMUs do not get measurements directly from the field; instead they use analogue values of current and voltage waveforms provided by CTs and PTs, respectively. These values are time-stamped by PMUs to an absolute reference provided by GPS in order to generate the sampled current and voltage phasors [16]. The waveforms provided by the instrument transformers have errors in both magnitude and phase, but the error in phase can be accurately compensated and calibrated out using digital signal processing (DSP) techniques [100]. The errors in magnitude of the waveforms provided by the instrument transformers are limited by the accuracy class of these instruments. There are two main standards according to which instrument transformers are designed: IEC 60044 [101] and IEEE C57.13 [102]. Both of these standards specify accuracies in the range of

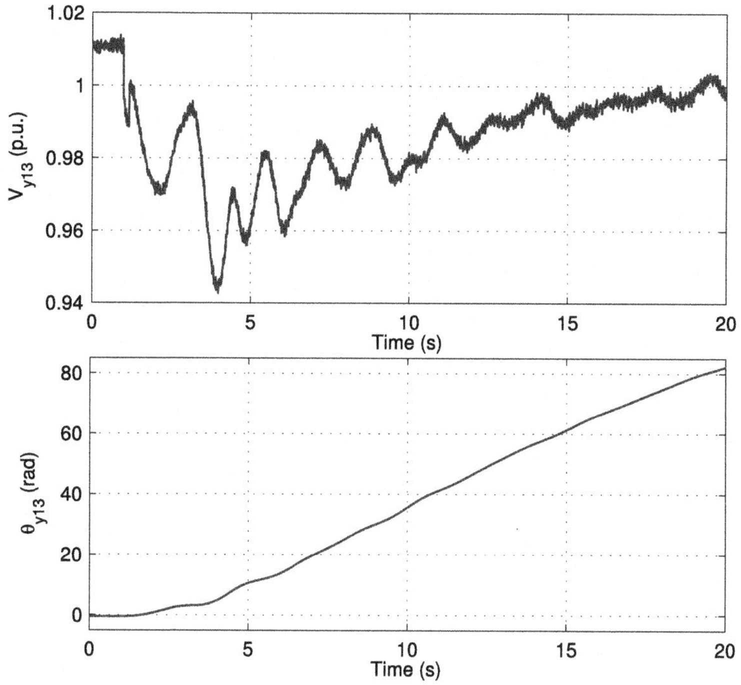

Figure 4.4 Generated measurements for V and θ for the 13th generation unit.

±0.1% to ±0.3% for the measurement of voltage and current magnitudes using modern CTs and PTs.

Thus, noises in the generated phase measurements were assumed to have a standard deviation of 0.1 mrad (or 10^{-4} rad), while noises in the generated magnitude measurements were assumed to have a standard deviation of 0.1% (or 10^{-3} p.u.), and hence $\sigma_{\theta_{w0}} = \sigma_{\phi_{w0}} = 10^{-4}$ rad and $\sigma_{V_{w0}} = \sigma_{I_{w0}} = 10^{-3}$ p.u. The "0" in $\sigma_{V_{w0}}$, $\sigma_{\theta_{w0}}$, $\sigma_{I_{w0}}$, and $\sigma_{\phi_{w0}}$ denotes that these are base case values. The variances for the generated noises for all the units were made equal to the base case values, and hence $\boldsymbol{P}_{zi} = \boldsymbol{P}_{z0} = \text{diag}\{10^{-6}, 10^{-8}\}$ and $\boldsymbol{P}_{wi} = \boldsymbol{P}_{w0} = \text{diag}\{10^{-6}, 10^{-8}\}$, $i = 1, 2, \ldots, 16$, from (4.43), (4.50).

White Gaussian noises with aforementioned variances were added to the sampled values of the actual signals in order to generate measurements. Fig. 4.4 shows the generated V_y and θ_y for the 13th unit.

4.5.1.2 Process noise

Process noise needs to be included in a model due to modeling approximations and model integration errors. It is not as straightforward to find process noise variances as it is to find measurement noise variances. This is because it is difficult to obtain errors due to unmodeled dynamics, modeling approximations, and parameter uncertainties and to combine them with integration errors of the discrete model used by the state estimator. A practical and robust way of finding process noise variances is to estimate them using a perturbation observer [105], but this method is not required for finding process noise variances in the case study.

In the case study, as a power system is simulated using known subtransient DAEs and discrete forms of the same DAEs are used by the estimator, modeling errors are absent. The only errors which are present are due to the discretization of the DAEs in the state estimator. As the DAEs are discretized according to the first-order Euler approximation, the discretization error in state x is $T_0^2\ddot{x}/2$ (also known as local truncation error of Euler's method [106]), where T_0 is the step size or sampling period of discretization. Noting that $\ddot{x} \approx \Delta(\Delta x)/T_0^2$, the standard deviation of the process noise in $x(k)$ is taken to be $max\{|\Delta(\Delta x(k))|\}/2$, excluding any sudden changes during faults or other such disturbances.

Here $\Delta x(k) = x(k) - x(k-1)$, $\Delta(\Delta x(k)) = \Delta x(k) - \Delta x(k-1) = x(k) - 2x(k-1) + x(k-2)$, $3 \leq k \leq N$, and N is the total number of samples for which the system is simulated (for example, for a 15-s simulation $N = (15\text{ s}) \times (120\text{ Hz}) = 1800$). This expression for process noise variance may be better understood with an example. Fig. 4.5 shows the state changes in δ and ω for the 13th unit.

It can be observed from the figure that $max\{|\Delta(\Delta\delta_{13}(k))|\}/2$ is 6×10^{-4} and $max\{|\Delta(\Delta\omega_{13}(k))|\}/2$ is 6×10^{-6}. Hence variances of noises in δ_{13} and ω_{13} are taken as 3.6×10^{-7} and 3.6×10^{-11}, respectively. This technique is used to find $\boldsymbol{P}_{vi}$ for all the machines. For the three different types of machines in the system, $\boldsymbol{P}_{vi}$ is found to be

$$\boldsymbol{P}_{vi} = \text{diag}\{1.6 \times 10^{-7}, 1.6 \times 10^{-11}, 4 \times 10^{-10}, 4 \times 10^{-10}, 9 \times 10^{-10}, 2.5 \times 10^{-9}, 3.6 \times 10^{-9}, 4 \times 10^{-6}, 2.5 \times 10^{-7}\},\ i = 1, 2, \ldots, 8, \quad (4.60)$$

$$\boldsymbol{P}_{v9} = \text{diag}\{1.6 \times 10^{-7}, 1.6 \times 10^{-11}, 4 \times 10^{-8}, 4 \times 10^{-10}, 9 \times 10^{-10}, 2.5 \times 10^{-9}, 3.6 \times 10^{-9}, 1 \times 10^{-12}, 2.5 \times 10^{-9}, 1.6 \times 10^{-9}\}, \quad (4.61)$$

$$\boldsymbol{P}_{vi} = \text{diag}\{3.6 \times 10^{-7}, 3.6 \times 10^{-11}, 6.4 \times 10^{-11}, 1.6 \times 10^{-9}, 2.5 \times 10^{-9}, 9 \times 10^{-10}\},\ i = 10, 11, \ldots, 16. \quad (4.62)$$

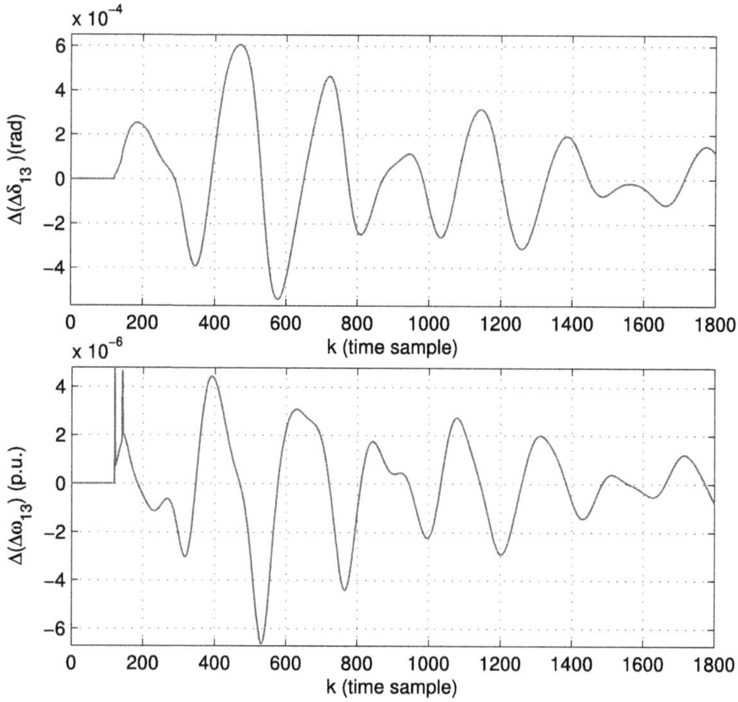

Figure 4.5 State changes in δ and ω for the 13th generation unit.

4.5.2 Simulation results and discussion

In the start of the simulation, the system is operating in a steady condition. Then at $t = 1$ s, a disturbance is created by a three-phase fault and the fault is cleared after 0.18 s by outage (or opening) of one of the tie lines between buses 53–54. The ith decentralized UKF algorithm, as given in Section 4.4, is running along with the simulation of the ith unit. The generated measurements from each unit are given as input to the corresponding UKF. The simulated states, along with their real-time estimated values, have been plotted for each unit. Corresponding estimation errors for various states have also been plotted. Due to space constraints, plots for only three units (of different types) have been shown: unit 3 of type 1 (Fig. B.1, Fig. B.3, and Fig. B.5 and corresponding errors in Fig. B.2, Fig. B.4, and Fig. B.6, respectively), unit 9 of type 2 (Fig. B.7, Fig. B.9, and Fig. B.11 and corresponding errors in Fig. B.8, Fig. B.10, and Fig. B.12, respectively), and

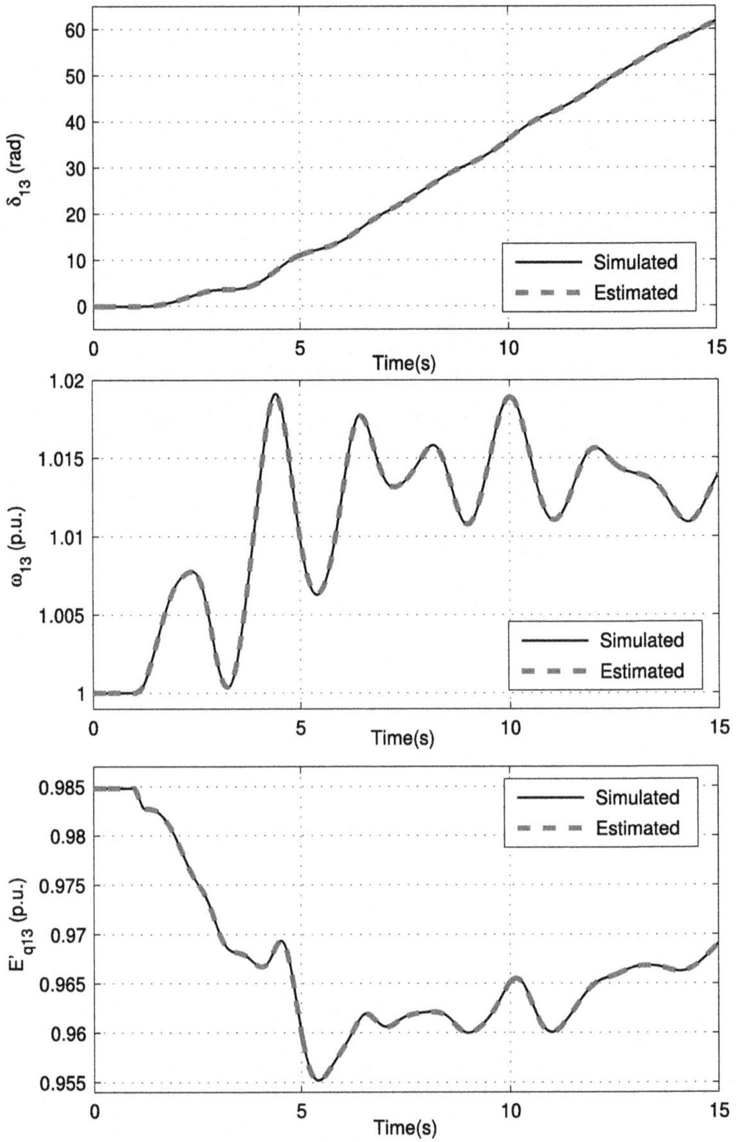

Figure 4.6 Estimated vs. simulated values for δ, ω, and E_q' of the 13th unit.

unit 13 of type 3 (Fig. 4.6 and Fig. 4.8 and corresponding errors in Fig. 4.7 and Fig. 4.9, respectively). The plots for unit 3 and unit 9 are given in Appendix B, while the plots for unit 13 are given in this chapter.

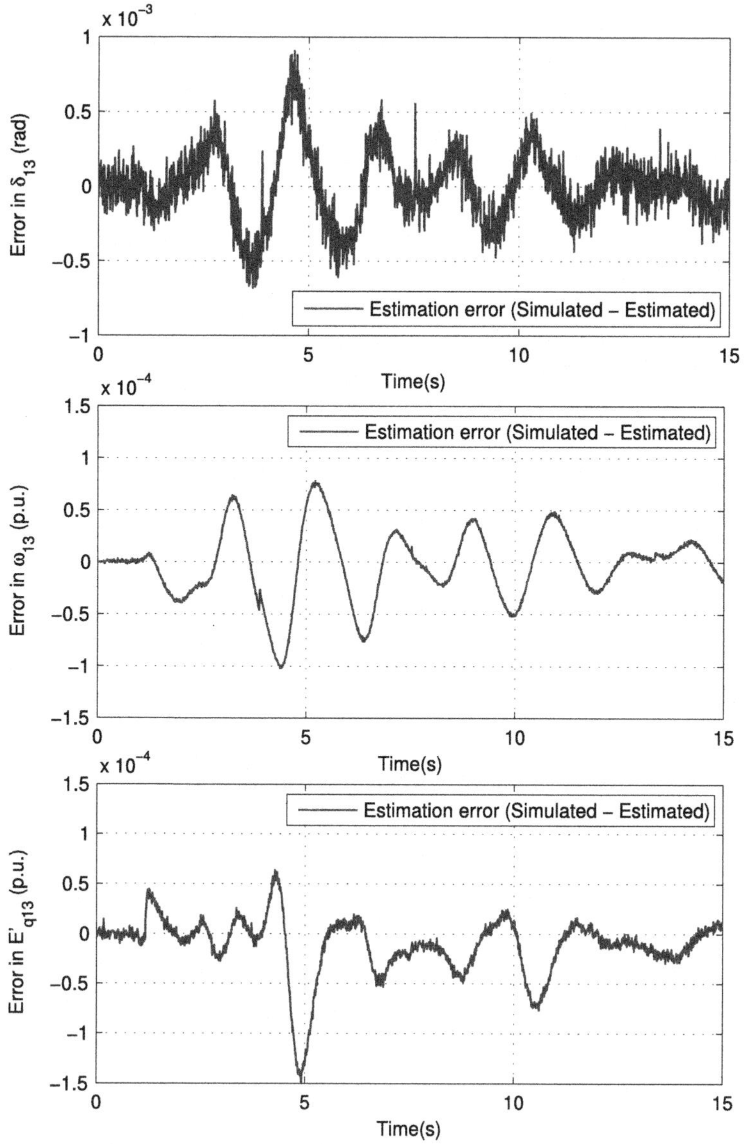

Figure 4.7 Estimation errors for δ, ω, and E_q' of the 13th unit.

4.5.2.1 Estimation accuracy

It can be seen in Figs. B.1–B.12 and Figs. 4.6–4.9 that for every dynamic state, the plot of estimated values almost coincides with those of the sim-

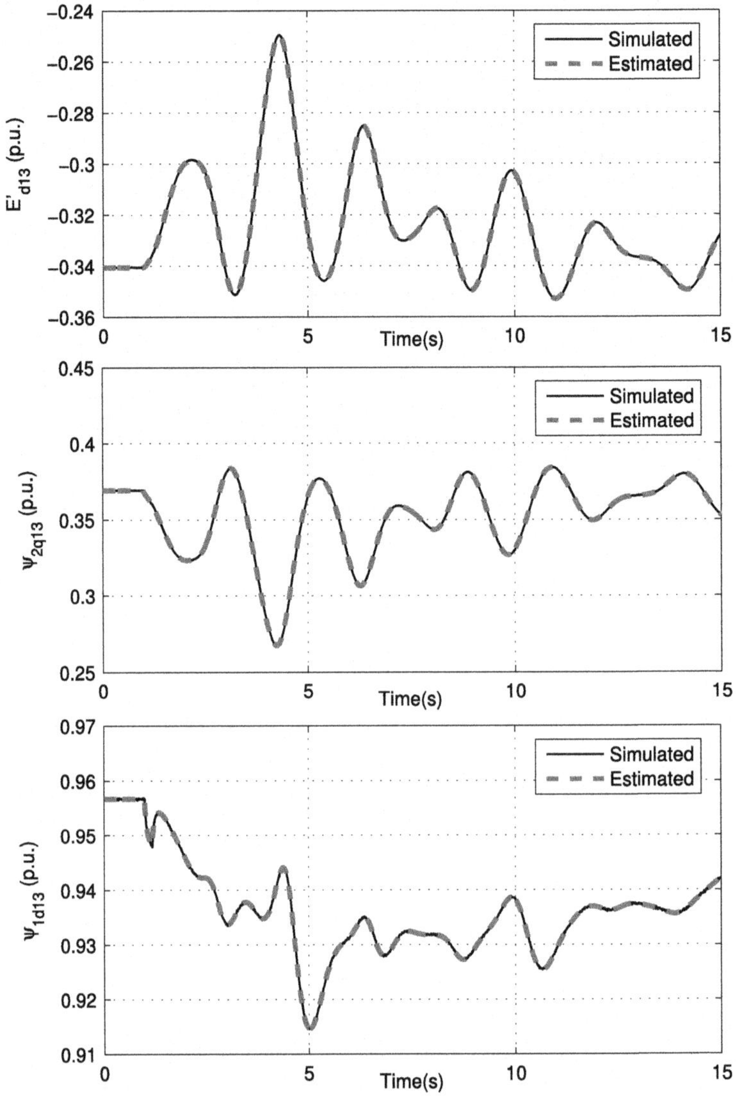

Figure 4.8 Estimated vs. simulated values for E'_d, Ψ_{2q}, and Ψ_{1d} of the 13th unit.

ulated values and the maximum estimation error in a state remains within 2% of the maximum deviation in the state (not considering the errors during and just after a disturbance). Thus, it is evident that the decentralized UKF scheme generates accurate estimates of all the dynamic states of a

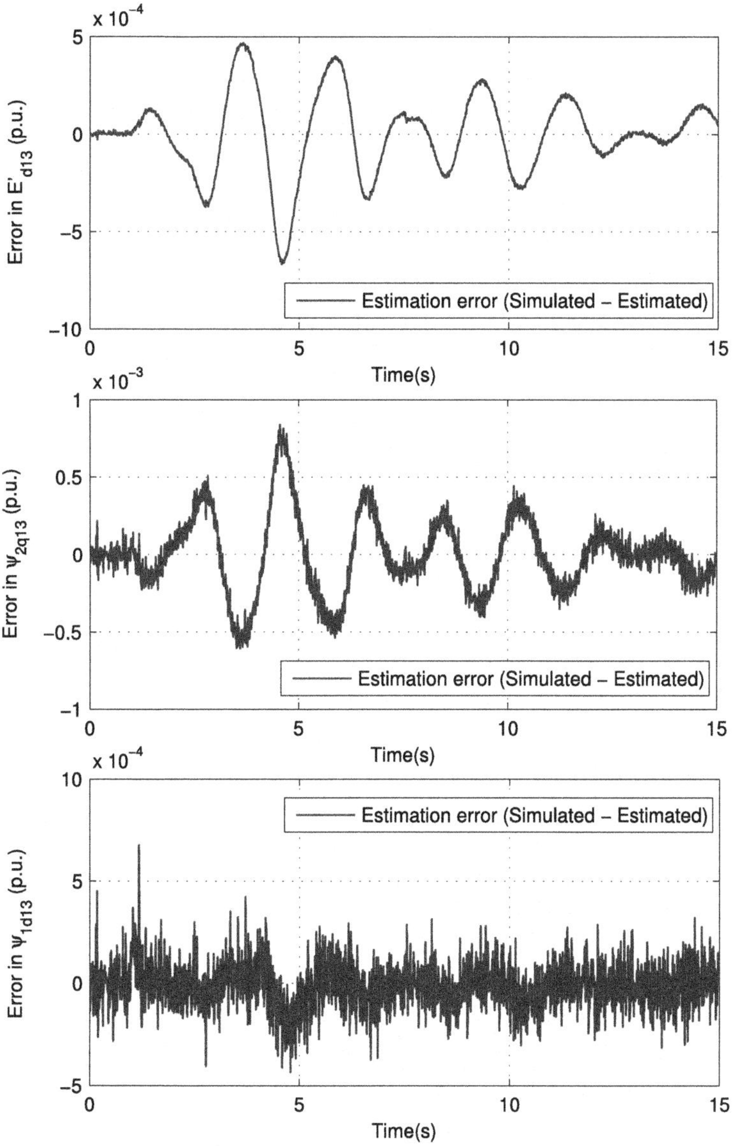

Figure 4.9 Estimation errors for E_d', Ψ_{2q}, and Ψ_{1d} of the 13th unit.

generating unit. As all the generator states have been estimated with high accuracies, they can be reliably used for further control and security decisions.

Table 4.1 Comparison of computational speeds

	Average computational time for one iteration (in ms)	
Test system	**Decentralized algorithm**	**Centralized algorithm**
IEEE 30-bus	0.33	1.45
IEEE 68-bus	0.33	12.4
IEEE 145-bus	0.33	139

4.5.2.2 Computational feasibility

The DSE-with-PMU algorithm has been tested on two more standard IEEE test systems to assess its scalability. As the measurements are updated every 8.33 ms ($T_0 = 8.33$ ms), a single iteration of the algorithm should not require more than 8.33 ms, otherwise the algorithm would not run in real-time. The average time for one iteration has been tabulated in Table 4.1 for the three test systems. A centralized scheme for DSE (given in [30]) is also implemented on all the test systems, and the corresponding average iteration times have been tabulated in Table 4.1. It can be inferred from Table 4.1 that the computational speed of the decentralized DSE-with-PMU algorithm is very high and it remains independent of the size of the system, while the centralized algorithm becomes slow and infeasible for large systems (68-bus and 145-bus systems).

4.5.2.3 Sensitivity to noise

The robustness of the decentralized DSE-with-PMU algorithm to higher noise variances is also tested. For this, the variances $\boldsymbol{P}_{zi}$ and $\boldsymbol{P}_{wi}$ were varied in multiples of tens of their base case values, $\boldsymbol{P}_{z0}$ and $\boldsymbol{P}_{w0}$, and the effect on estimation accuracy is observed. Fig. 4.10 shows the effect of variations in noise variances on the estimation of ω for the *type 2* of the generation unit, and Fig. 4.11 shows the corresponding estimation errors. The plots have been shown for a portion of the total simulation time for clarity.

It is evident from Fig. 4.10 and Fig. 4.11 that the algorithm is robust, with minor errors in estimated values, even when the noise variances are a hundred times their base case values. When the noise variances are a thousand times the base case variances, the estimated states have significant errors and deviations and hence become unusable.

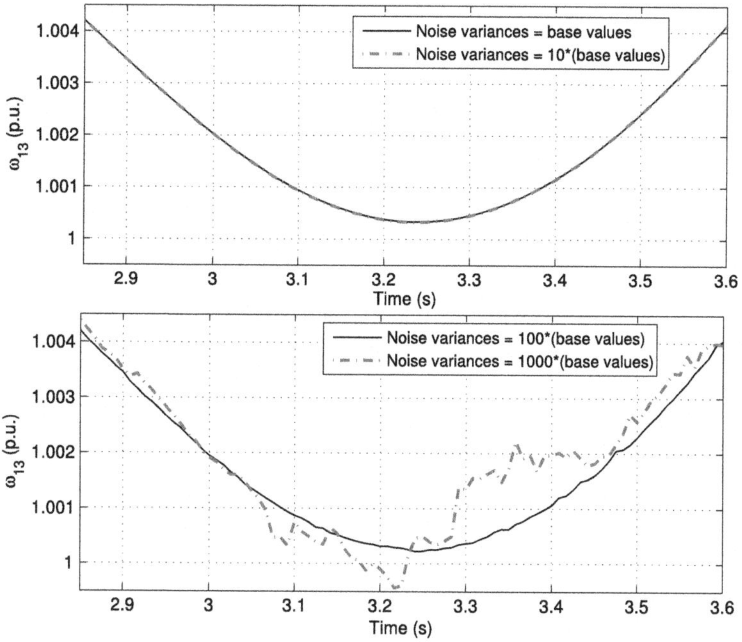

Figure 4.10 Effect of noise variances on the accuracy of estimation.

4.6 BAD-DATA DETECTION

PMU signals not only suffer from noise, but they are also prone to gross errors; and therefore a bad-data detection algorithm is required for the decentralized DSE-with-PMU estimator. Bad-data detection in UKF is based on the fact that the ratio between the deviation of actual measurement from the predicted measurement and the expected standard deviation of the predicted measurement remains bounded in a narrow band in the absence of any bad data; this ratio is called normalized innovation ratio [30,31]. Mathematically, this fact may be stated using (4.63) and (4.64), where $\lambda_{y_{i,1}}$ and $\lambda_{y_{i,2}}$ are the normalized innovation ratios for the two measurements $y_{i,1} = I_{yi}$ and $y_{i,2} = \phi_{yi}$, respectively (recall that $\boldsymbol{y}_i = [I_{yi}, \phi_{yi}]^T$), $\hat{\boldsymbol{y}}_i^- = [\hat{y}_{i,1}^-, \hat{y}_{i,2}^-]^T$, $P_{yi,1}^-$ is the first diagonal element of $\boldsymbol{P}_{yi}^-$, and $P_{yi,2}^-$ is the second diagonal element of $\boldsymbol{P}_{yi}^-$. We have

$$\lambda_{y_{i,1}} < \lambda_0, \text{ where } \lambda_{y_{i,1}} = \frac{|y_{i,1} - \hat{y}_{i,1}^-|}{\sqrt{P_{yi,1}^-}}, \tag{4.63}$$

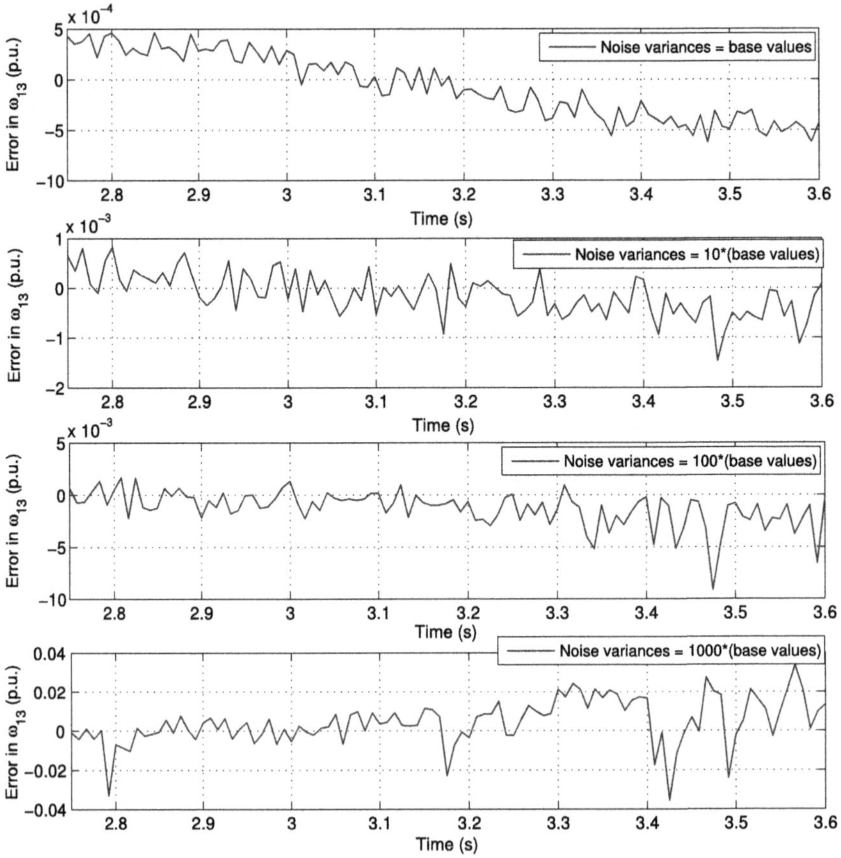

Figure 4.11 Effect of noise variances on estimation errors.

$$\lambda_{y_{i,2}} < \lambda_0, \text{ where } \lambda_{y_{i,2}} = \frac{|y_{i,2} - \hat{y}_{i,2}^{-}|}{\sqrt{P_{yi,2}^{-}}}; \tag{4.64}$$

λ_0 depends on the type of the system, and it may be found using offline simulations [30,31]. For the system in this case study, λ_0 is found to be 10. Hence, a measurement is labeled as a bad measurement if its normalized innovation ratio turns out to be more than λ_0 in a given sample; it is then discarded and the actual measurement is assumed to be the same as the predicted measurement for that sample.

The above technique for bad-data detection would have worked flawlessly if there were no bad data present in the states or input. But since pseudoinputs are used in the decentralized UKF algorithm, which are in reality measurements, bad data may also be present in these pseudoinputs. Innovation ratios are not defined for pseudoinputs, and hence we cannot directly detect bad data in them; but an indirect method may be used to do so. This method is based on the fact that the predicted measurements are influenced by bad data in the pseudoinputs but the actual measurements remain independent of these bad data, and hence in the case of bad data in pseudoinputs no correlation exists between the actual measurements and the predicted measurements. In other words, if bad data are introduced in one or more pseudoinput(s) in a given sample, then both $\hat{\boldsymbol{y}}_i^-$ and $\boldsymbol{P}_{yi}^-$ would change significantly from their correct values, and this change will be completely uncorrelated with $\boldsymbol{y}_i$, even if bad data are present in $\boldsymbol{y}_i$ as well (assuming that all the bad data are introduced randomly and independently), and thus the values of both $\lambda_{y_{i,1}}$ and $\lambda_{y_{i,2}}$ are expected to exceed λ_0 in such an event. Thus, we need to modify the aforementioned technique of bad-data detection and discard all the pseudoinputs if both $\lambda_{y_{i,1}}$ and $\lambda_{y_{i,2}}$ exceed λ_0 in a given sample, and use the latest uncorrupted pseudoinputs instead. Thus, the bad-data detection for the kth sample takes place according to the following algorithm and Fig. 4.12.

A bad-data detector based on **Algorithm 4.2** is implemented and integrated in the decentralized UKF algorithm. Values of $\lambda_{y_{i,1}}$ and $\lambda_{y_{i,2}}$ and the estimated rotor velocity for $i = 13$ are shown for three cases, all in Fig. 4.13. We distinguish between the following scenarios:

1. *Bad data present only in one of the measurements:* In this case bad data are introduced in the measurement ϕ_{y13} of magnitude +0.4 p.u. (i.e., the measured value of ϕ_{y13} is 0.01 p.u. above its true value), at time $t = 5$ s. It may be observed that the bad-data detector effectively handles this anomaly, there is no effect on $\lambda_{y_{13,1}}$, and the estimation process remains unaffected.
2. *Bad data present only in one of the pseudoinputs:* In this case bad data are introduced in the pseudoinput V_{y13} of magnitude +0.2, at time $t = 10$ s. It may be observed that both $\lambda_{y_{13,1}}$ and $\lambda_{y_{13,2}}$ become unbounded, but the bad-data detector effectively handles this anomaly as well, as the estimation process remains unaffected.
3. *Bad data present simultaneously in one of the measurements and in one of the pseudoinputs:* In this case bad data are introduced in the measurement ϕ_{y13} of magnitude +0.4 p.u., and other bad data are introduced in the pseudoinput V_{y13} of magnitude +0.2 p.u., both at $t = 15$ s. It may be observed that both $\lambda_{y_{13,1}}$ and $\lambda_{y_{13,2}}$ become unbounded, as in the

Algorithm 4.2 Bad-data detection in DSE for the ith generation unit.

STEP 1:

Perform the first four steps of **Algorithm 4.1**.

STEP 2:

Find $\lambda_{y_{i,1}}$ and $\lambda_{y_{i,2}}$ according to (4.63) and (4.64), respectively.

STEP 3:

if $\lambda_{y_{i,1}} < \lambda_0$ and $\lambda_{y_{i,2}} < \lambda_0$ **then** goto **STEP 5**

else

if $\lambda_{y_{i,1}} > \lambda_0$ and $\lambda_{y_{i,2}} < \lambda_0$ **then** $y_{i,1} = \hat{y}_{i,1}^-$, goto **STEP 5**

else

if $\lambda_{y_{i,1}} < \lambda_0$ and $\lambda_{y_{i,2}} > \lambda_0$ **then** $y_{i,2} = \hat{y}_{i,2}^-$, goto **STEP 5**

else

if $\lambda_{y_{i,1}} > \lambda_0$ and $\lambda_{y_{i,2}} > \lambda_0$ **then** discard $\boldsymbol{u}_i'$ and again perform the first four steps of **Algorithm 4.1** using the latest uncorrupted value of $\boldsymbol{u}_i'$. Again find $\lambda_{y_{i,1}}$ and $\lambda_{y_{i,2}}$ according to (4.63) and (4.64), respectively.

STEP 4:

if $\lambda_{y_{i,1}} < \lambda_0$ and $\lambda_{y_{i,2}} < \lambda_0$ **then** goto **STEP 5**

else

if $\lambda_{y_{i,1}} > \lambda_0$ and $\lambda_{y_{i,2}} < \lambda_0$ **then** $y_{i,1} = \hat{y}_{i,1}^-$, goto **STEP 5**

else

if $\lambda_{y_{i,1}} < \lambda_0$ and $\lambda_{y_{i,2}} > \lambda_0$ **then** $y_{i,2} = \hat{y}_{i,2}^-$, goto **STEP 5**

else

if $\lambda_{y_{i,1}} > \lambda_0$ and $\lambda_{y_{i,2}} > \lambda_0$ **then** $y_{i,1} = \hat{y}_{i,1}^-$ and $y_{i,2} = \hat{y}_{i,2}^-$.

STEP 5:

Perform the last two steps of **Algorithm 4.1**.

previous case, and the bad-data detector effectively handles this anomaly as well.

Thus, the two-stage bad-data detector successfully filters out bad data in all the three possible cases.

4.7 OTHER PMU-BASED METHODS OF DSE

In the last five years, considerable research has been undertaken to explore various aspects of DSE – accuracy, noise resilience, efficiency, and stability – and several methods of both centralized and decentralized DSE have been proposed [32–41]. Even though the various decentralized DSE methods use different techniques of signal processing for finding the state estimates

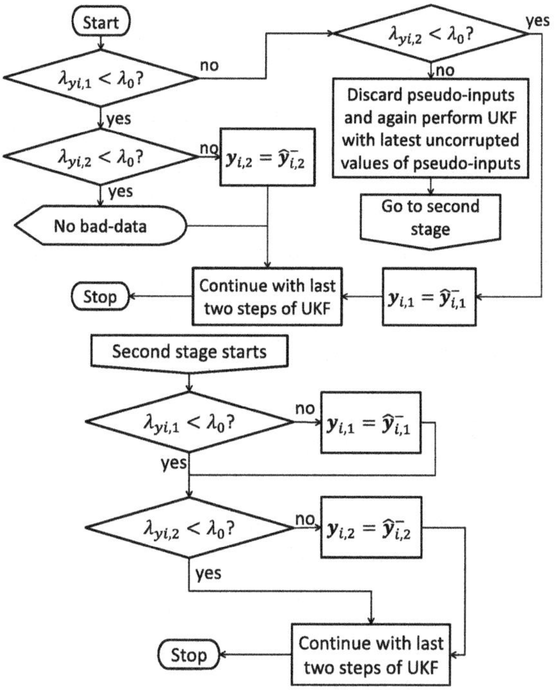

Figure 4.12 Flowchart for bad-data detection.

(such as extended Kalman filter, UKF, particle filter, ensemble Kalman filter, etc.), the basic underlying principle of decentralization in all of these methods remains the same. That is, as discussed in this chapter, first, the concept of pseudoinputs is applied to decouple the dynamic model of a power system at the point of coupling of each generating unit with the rest of the system; second, a standard filtering technique is applied to the PMU measurements acquired at one of the points of coupling, in conjugation with the known parameters of the corresponding generating unit, to obtain the final dynamic state estimates of the generating unit. A brief description and a comparative study of the filtering techniques used in PMU-based decentralized DSE is given in [39].

4.8 SUMMARY

A scheme for decentralized estimation of the dynamic states of a power system has been presented in this chapter. The scheme preserves nonlinearity

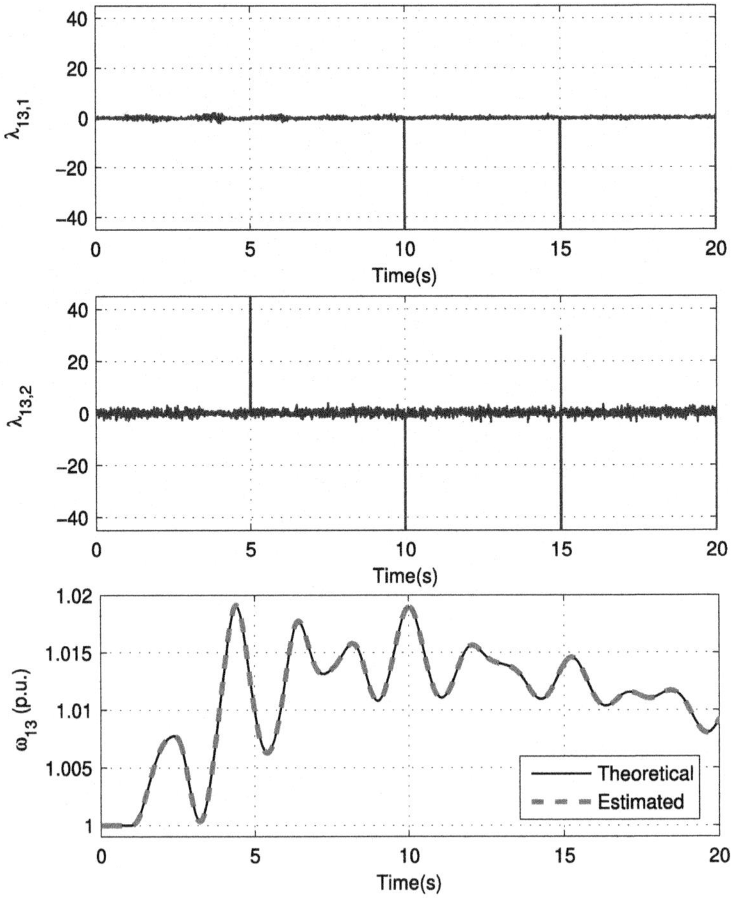

Figure 4.13 Bad-data detection.

in the system dynamics through the application of UKF. The basic idea of decentralization in the scheme is based on treating some of the measured signals as pseudoinputs. The advantages of the decentralized DSE-with-PMU scheme over the centralized schemes have been presented in terms of speed, feasibility, simplicity, and accuracy. The scheme is also robust to moderately high noise levels and gross errors in measurement signals.

The key advantages of the presented scheme may be summarized as follows:

1. The signals required for estimation (which are the generator voltage and current) are easy to measure using PMUs.

2. Each distributed estimator has to estimate only local states of the corresponding generation unit. Therefore the estimator is very fast and its speed remains independent of the size of the system, unlike a centralized scheme. This is the biggest advantage over a centralized scheme as for a large power system, the number of states a central estimator has to estimate is very large, requiring huge computational capacity for real-time estimation.
3. Remote signals need not be transmitted; therefore the estimation process is not affected by network problems such as transmission delays and losses. Also, the signal sampling rates are not limited by network bandwidth.
4. State estimation for one generation unit is completely independent from estimation for other units. Thus, errors in estimation remain isolated and are easier to pinpoint than in a centralized estimation scheme.
5. PMUs only need to be installed at each generation unit, and most power stations are likely to have PMUs installed.

CHAPTER 5

Dynamic Parameter Estimation of Analogue Voltage and Current Signals

In this chapter, we present a decentralized estimation of the parameters of sinusoidal signals of voltage and current measured at a bus in a power system. The analogue sinusoidal signals of current and voltage are measured using current/voltage transformers (CT/VTs) installed at a node or bus in the power network. At a given instant of time, each of these signals can be described using some attributes or parameters such as the magnitude, phase, and frequency of the signal. These parameters, estimated in real-time, can be used for monitoring and protecting the generating unit, transmission line(s), load(s), and any other device that is connected to the bus. Another application is to perform dynamic state estimation for a generator (as described in Chapter 6).

Several methods have been proposed in the literature for estimating the parameters of a sinusoidal signal, but most of these methods are computationally expensive and, hence, are not suitable for real-time applications [107]. Recently, an interpolated discrete-time Fourier transform (DFT)-based estimation method was proposed in [108] and has been shown to be both fast and accurate enough for real-time control applications in power systems. This method is studied and further developed in this chapter for finding the estimates of frequency, magnitude, and phase of the fundamental components of measurements obtained from the CTs and VTs.

Section 5.1 presents the theory of interpolated DFT and derives the expressions for the mean values of the parameter estimates. Section 5.2 presents the theory of Cramer–Rao bounds and uses it to derive the expressions for the variance of the parameter estimates. Section 5.3 presents an implementation example for the interpolated DFT estimator, while Section 5.4 summarizes the chapter.

5.1 INTERPOLATED DFT-BASED ESTIMATION

The fundamental component of a sinusoidal signal can be extracted by multiplying the signal with a suitable window function which eliminates

Dynamic Estimation and Control of Power Systems
https://doi.org/10.1016/B978-0-12-814005-5.00016-9

other harmonics and higher-frequency components in the signal, followed by finding its DFT. One such function is the Hanning window function given by $h_k = \sin^2(\frac{\pi k}{N})$, and if this function is multiplied with N samples of an analogue signal $Y(t)$ sampled at a frequency f_s, then the DFT of the product is given by $Z(\lambda)$ as follows [108]:

$$\begin{aligned} Z(\lambda) &= \sum_{k=0}^{N-1} Y_k h_k e^{-\frac{j2\pi k\lambda}{N}} \\ &= \frac{Y_m}{2j} e^{j\theta} W(\lambda - \frac{fN}{f_s}) - \frac{Y_m}{2j} e^{-j\theta} W(\lambda + \frac{fN}{f_s}), \end{aligned} \tag{5.1}$$

where Y_m, θ, and f are the magnitude, phase, and frequency of Y's fundamental component, respectively, $\lambda \in \{0, 1, ..., N-1\}$, and $W(\lambda)$ is the following DFT of the Hanning window function:

$$W(\lambda) = \sum_{k=0}^{N-1} h_k e^{-\frac{j2\pi k\lambda}{N}} = \sum_{k=0}^{N-1} \sin^2(\frac{\pi k}{N}) e^{-\frac{j2\pi k\lambda}{N}}. \tag{5.2}$$

The key concept in interpolated DFT-based estimation is to approximate $W(\lambda)$ with the following expression, provided that $N >> 1$ and $\lambda << N$ [108,109]:

$$W(\lambda) \approx \frac{N}{4\pi j} \frac{(1 - e^{-j2\pi\lambda})}{(\lambda - \lambda^3)}. \tag{5.3}$$

By substituting (5.3) in (5.1), $Z(\lambda)$ can be expressed as follows for $N >> 1$ and $\lambda << N$:

$$\begin{aligned} Z_\lambda = Z(\lambda) = \frac{\hat{Y}_m N}{8\pi} \Bigg[& \frac{e^{j\hat{\theta}} (e^{-j2\pi(\lambda - \frac{\hat{f}N}{f_s})} - 1)}{(\lambda - \frac{\hat{f}N}{f_s}) - (\lambda - \frac{\hat{f}N}{f_s})^3} \\ & - \frac{e^{-j\hat{\theta}} (e^{-j2\pi(\lambda + \frac{\hat{f}N}{f_s})} - 1)}{(\lambda + \frac{\hat{f}N}{f_s}) - (\lambda + \frac{\hat{f}N}{f_s})^3} \Bigg], \end{aligned} \tag{5.4}$$

where $\hat{Y}_m$, $\hat{\theta}$, and $\hat{f}$ denote the estimates of Y_m, θ, and f, respectively. As (5.4) has three unknowns (which are $\hat{Y}_m$, $\hat{\theta}$, and $\hat{f}$), three distinct equations are required to estimate these unknowns. This can be done by choosing any three distinct values of λ in (5.4) (say $\lambda = 1$, $\lambda = 2$, and $\lambda = 3$). The obtained values of $\hat{Y}_m$, $\hat{\theta}$, and $\hat{f}$ will have associated estimation errors which will depend on N and on the values of λ which are used for generating the three distinct equations. More precisely, these estimation errors are inversely proportional to N^4 [108], and, hence, N should be as large as practically feasible. In this chapter N is taken to be in the order of 10^3, as this is the highest order for N for which interpolated DFT can run on a state-of-the-art DSP processor without overloading it [108] (the DSP processor used in [108] is a TMS320C6713 with a 225-MHz clock rate and a 264-kB on-chip RAM; overloading refers to overall processor usage of above 95%). Also, for a given N, the estimation errors are minimized if the choices for λ are taken as $\lambda = 0$, $\lambda = 1$, and $\lambda = 2$, provided that $\frac{\hat{f}N}{f_s} < 2.1$; otherwise, for $2.1 < \frac{\hat{f}N}{f_s} < 3$, the errors are minimized if the choices are $\lambda = 1$, $\lambda = 2$, and $\lambda = 3$ [108]. The value of $\frac{\hat{f}N}{f_s}$ should not be greater than 3 as otherwise the delay in obtaining the estimated values becomes too large (that is, more than two cycles, or more than 0.04 s for a 50-Hz power system), and at the same time it should not be too small as in that case the accuracy of the estimation is diminished [108]. In this chapter an intermediate value of $\frac{\hat{f}N}{f_s} \approx 1.5$ has been taken and, hence, the choices of $\lambda = 0$, $\lambda = 1$, and $\lambda = 2$ are applicable.

Remark 5.1. The fraction $\frac{\hat{f}N}{f_s}$ is an unknown quantity as f needs to be estimated. But because of power system operational requirements [2], f should remain within 5% of the base system frequency, f_0 (which is usually 50 Hz or 60 Hz), and, hence, if N and f_s are chosen such that $\frac{f_0 N}{f_s} = 1.5$, then $\frac{\hat{f}N}{f_s} \approx 1.5$. It should be noted that the interpolated DFT-based estimation remains fully functional even if f transiently violates the operational limit of being within 5% of f_0, although this increases the error of estimation.

5.1.1 Expressions for mean values of the parameter estimates

The three equations which are obtained by putting $\lambda = 0$, $\lambda = 1$, and $\lambda = 2$ in (5.4) can be written in matrix form as follows:

$$\begin{bmatrix} \frac{\frac{\hat{f}N}{f_s}-2}{\frac{\hat{f}N}{f_s}+1} & \frac{\frac{\hat{f}N}{f_s}+2}{\frac{\hat{f}N}{f_s}-1} & Z_0 \\ 1 & 1 & Z_1 \\ \frac{\frac{\hat{f}N}{f_s}}{\frac{\hat{f}N}{f_s}-3} & \frac{\frac{\hat{f}N}{f_s}}{\frac{\hat{f}N}{f_s}+3} & Z_2 \end{bmatrix} \begin{bmatrix} \frac{\hat{Y}_m N e^{j\hat{\theta}}\left(e^{\frac{j2\pi\hat{f}N}{f_s}}-1\right)}{8\pi\frac{\hat{f}N}{f_s}\left(\frac{\hat{f}N}{f_s}-1\right)\left(\frac{\hat{f}N}{f_s}-2\right)} \\ \frac{\hat{Y}_m N e^{-j\hat{\theta}}\left(e^{-\frac{j2\pi\hat{f}N}{f_s}}-1\right)}{8\pi\frac{\hat{f}N}{f_s}\left(\frac{\hat{f}N}{f_s}+1\right)\left(\frac{\hat{f}N}{f_s}+2\right)} \\ -1 \end{bmatrix} = \begin{bmatrix} 0 \\ 0 \\ 0 \end{bmatrix}. \tag{5.5}$$

Eq. (5.5) implies that the product of a square matrix and a column vector is equal to a zero vector, despite both the matrix and the vector having nonzero elements. This can only happen if the columns of the matrix are linearly dependent, that is, the determinant of the matrix is zero, given as follows:

$$\begin{vmatrix} \frac{\hat{f}N-2f_s}{\hat{f}N+f_s} & \frac{\hat{f}N+2f_s}{\hat{f}N-f_s} & Z_0 \\ 1 & 1 & Z_1 \\ \frac{\hat{f}N}{\hat{f}N-3f_s} & \frac{\hat{f}N}{\hat{f}N+3f_s} & Z_2 \end{vmatrix} = 0. \tag{5.6}$$

Simplification of the above determinant gives $\hat{f}$ as follows:

$$\hat{f} = \frac{f_s}{N}\sqrt{\frac{Z_0+2Z_1+9Z_2}{Z_0-2Z_1+Z_2}}, \tag{5.7}$$

where $\hat{\theta}$ can be obtained by substituting the above value of $\hat{f}$ back into (5.4) and eliminating $\hat{Y}_m$. To do this, the equation which is obtained by putting $\lambda = 0$ in (5.4) is divided by the equation obtained by putting $\lambda = 1$ in (5.4). Then we have

$$\frac{Z_0}{Z_1} = \frac{e^{j\hat{\theta}}B + e^{-j\hat{\theta}}C}{e^{j\hat{\theta}}E + e^{-j\hat{\theta}}F}, \quad B = \frac{1-e^{\frac{j2\pi\hat{f}N}{f_s}}}{\frac{\hat{f}N}{f_s} - [\frac{\hat{f}N}{f_s}]^3}, \quad C = \frac{1-e^{-\frac{j2\pi\hat{f}N}{f_s}}}{\frac{\hat{f}N}{f_s} - [\frac{\hat{f}N}{f_s}]^3},$$

$$E = \frac{1-e^{\frac{j2\pi\hat{f}N}{f_s}}}{\frac{\hat{f}N}{f_s} - 1 - [\frac{\hat{f}N}{f_s} - 1]^3}, \quad F = \frac{1-e^{-\frac{j2\pi\hat{f}N}{f_s}}}{\frac{\hat{f}N}{f_s} + 1 - [\frac{\hat{f}N}{f_s} + 1]^3}. \tag{5.8}$$

Solving for $e^{j\hat{\theta}}$ using (5.8) gives the following expression:

$$e^{j\hat{\theta}} = \sqrt{\frac{Z_0F - Z_1C}{Z_1B - Z_0E}} \Rightarrow \hat{\theta} = \frac{1}{2j}\ln\left\{\frac{Z_0F - Z_1C}{Z_1B - Z_0E}\right\}. \tag{5.9}$$

It should be noted that the value of $\hat{\theta}$ lies in the domain $(-\frac{\pi}{2}, \frac{\pi}{2}]$ using the definition of the principal value of a natural logarithm. Also, its value does not have a standard reference frame (unlike a phase value measured using PMU), and its reference depends on the internal clock of the DSP processor. Using (5.9) and (5.4) (with $\lambda = 0$), $\hat{Y}_m$ becomes

$$\hat{Y}_m = 8\pi Z_0 / \left[N\left\{Be^{j\hat{\theta}} + Ce^{-j\hat{\theta}}\right\}\right], \tag{5.10}$$

where B, C, and $e^{j\hat{\theta}}$ are given by (5.8)–(5.9).

It should be noted that $\hat{f}$, $\hat{\theta}$, and $\hat{Y}_m$ are real quantities, but they are obtained as functions of complex quantities (given in the right-hand sides (RHSs) of (5.7), (5.9), and (5.10), respectively). Hence, these quantities will have negligible but finite imaginary parts associated with them because of the finite computational accuracy of any computational device. Thus, the imaginary parts should be ignored and only the real parts of the RHSs should be assigned to $\hat{f}$, $\hat{\theta}$, or $\hat{Y}_m$. Also, as $\hat{f}$ and $\hat{Y}_m$ are strictly positive, absolute values of real parts of respective RHSs should be assigned to them.

5.2 VARIANCE OF PARAMETER ESTIMATES

It was found in [108] that the variance of the above estimate of $\hat{f}$ in (5.7) is approximately twice the minimum possible variance which is theoretically achievable using any unbiased estimator (known as Cramer–Rao bound [CRB] [110]). The CRB for frequency estimation of a sinusoidal signal has been derived in [110] and is given as follows:

$$\mathrm{CRB}(\hat{f}) = \left(\frac{f_s}{2\pi}\right)^2 \frac{24\sigma_Y^2}{\hat{Y}_m^2 N(N^2-1)}, \tag{5.11}$$

where σ_Y^2 is the variance of noise in Y (in p.u., with magnitude of Y as base, that is, if the noise is 3% of Y's magnitude, then $\sigma_Y^2 = (0.03)^2$). CRBs for $\hat{Y}_m$, $\hat{\theta}$, and $\hat{f}$ are derived as follows and are given by $\mathrm{CRB}(\hat{Y}_m)$ (in p.u.), $\mathrm{CRB}(\hat{\theta})$ (in rad^2), and $\mathrm{CRB}(\hat{f})$ (in Hz^2), respectively.

5.2.1 Cramer–Rao bounds for the parameters

Let the N samples of a sinusoidal signal Y, sampled at a sampling frequency of f_s, be given as follows:

$$Y_k = Y_m \sin\phi_k + \epsilon_k, \quad \phi_k = \frac{2\pi kf}{f_s} + \theta, \quad \forall k = 1, 2, \ldots, N. \tag{5.12}$$

Here, ϵ_k is the noise in Y_k, and the variance of ϵ_k is σ_Y^2. The set of parameters which need to be estimated for (5.12) is $\boldsymbol{\Theta} = \{\Theta_1, \Theta_2, \Theta_3\} = \{Y_m, \theta, f\}$. A lower bound on the variance of any unbiased estimator of $\boldsymbol{\Theta}$ is given by the CRB [111], which is found using the inverse of the information matrix, $\mathbb{I}(\boldsymbol{\Theta})$. For (5.12), the (i, j)th element of the matrix $\mathbb{I}(\boldsymbol{\Theta})$ is given as follows:

$$\mathbb{I}_{i,j}(\boldsymbol{\Theta}) = \frac{1}{\sigma_Y^2} \sum_{k=1}^{N} \frac{\partial Y_k}{\partial \Theta_i} \frac{\partial Y_k}{\partial \Theta_j}, \quad \forall i, j \in \{1, 2, 3\}. \tag{5.13}$$

After evaluating the partial derivatives in (5.13) using (5.12), various elements of $\mathbb{I}(\boldsymbol{\Theta})$ are given as follows:

$$\begin{aligned}
\mathbb{I}_{1,1}(\boldsymbol{\Theta}) &= \frac{1}{2\sigma_Y^2} \sum_{k=1}^{N} (1 - \cos(2\phi_k)), \\
\mathbb{I}_{1,2}(\boldsymbol{\Theta}) &= \mathbb{I}_{2,1}(\boldsymbol{\Theta}) = \frac{1}{2\sigma_Y^2} \sum_{k=1}^{N} Y_m \sin(2\phi_k), \\
\mathbb{I}_{1,3}(\boldsymbol{\Theta}) &= \mathbb{I}_{3,1}(\boldsymbol{\Theta}) = \frac{1}{2\sigma_Y^2} \sum_{k=1}^{N} Y_m \frac{2\pi k}{f_s} \sin(2\phi_k), \\
\mathbb{I}_{2,2}(\boldsymbol{\Theta}) &= \frac{1}{2\sigma_Y^2} \sum_{k=1}^{N} Y_m^2 (1 + \cos(2\phi_k)), \\
\mathbb{I}_{2,3}(\boldsymbol{\Theta}) &= \mathbb{I}_{3,2}(\boldsymbol{\Theta}) = \frac{1}{2\sigma_Y^2} \sum_{k=1}^{N} Y_m^2 \frac{2\pi k}{f_s} (1 + \cos(2\phi_k)), \\
\mathbb{I}_{3,3}(\boldsymbol{\Theta}) &= \frac{1}{2\sigma_Y^2} \sum_{k=1}^{N} Y_m^2 \left(\frac{2\pi k}{f_s}\right)^2 (1 + \cos(2\phi_k)),
\end{aligned} \tag{5.14}$$

where $\phi_k = \frac{2\pi kf}{f_s} + \theta$.

Since $N >> 1$, $f << f_s$, and $fN/f_s \approx 1.5$, as explained in Section 5.1, the above elements of $\mathbb{I}(\boldsymbol{\Theta})$ get simplified to the following expressions using basic rules of summation of trigonometric series:

$$\begin{aligned}
&\mathbb{I}_{1,1}(\boldsymbol{\Theta}) = \frac{1}{2\sigma_Y^2} N, \quad \mathbb{I}_{2,2}(\boldsymbol{\Theta}) = \frac{1}{2\sigma_Y^2} Y_m^2 N, \\
&\mathbb{I}_{1,2}(\boldsymbol{\Theta}) = \mathbb{I}_{2,1}(\boldsymbol{\Theta}) = \mathbb{I}_{1,3}(\boldsymbol{\Theta}) = \mathbb{I}_{3,1}(\boldsymbol{\Theta}) = 0, \\
&\mathbb{I}_{2,3}(\boldsymbol{\Theta}) = \mathbb{I}_{3,2}(\boldsymbol{\Theta}) = \frac{1}{2\sigma_Y^2} Y_m^2 \frac{2\pi}{f_s} \frac{N(N+1)}{2}, \\
&\mathbb{I}_{3,3}(\boldsymbol{\Theta}) = \frac{1}{2\sigma_Y^2} Y_m^2 \left(\frac{2\pi}{f_s}\right)^2 \frac{N(N+1)(2N+1)}{6}.
\end{aligned} \tag{5.15}$$

With $\mathbb{I}(\boldsymbol{\Theta})$ defined as in (5.15), various elements of its inverse, $\mathbb{I}^{-1}(\boldsymbol{\Theta})$, are obtained as follows:

$$
\begin{aligned}
&\mathbb{I}_{1,1}^{-1}(\boldsymbol{\Theta}) = \frac{2\sigma_Y^2}{N}, \quad \mathbb{I}_{2,2}^{-1}(\boldsymbol{\Theta}) = \frac{4\sigma_Y^2}{Y_m^2}\frac{(2N+1)}{N(N-1)},\\
&\mathbb{I}_{1,2}^{-1}(\boldsymbol{\Theta}) = \mathbb{I}_{2,1}^{-1}(\boldsymbol{\Theta}) = \mathbb{I}_{1,3}^{-1}(\boldsymbol{\Theta}) = \mathbb{I}_{3,1}^{-1}(\boldsymbol{\Theta}) = 0,\\
&\mathbb{I}_{2,3}^{-1}(\boldsymbol{\Theta}) = \mathbb{I}_{3,2}^{-1}(\boldsymbol{\Theta}) = \frac{-12\sigma_Y^2}{Y_m^2 N(N-1)}\frac{f_s}{2\pi},\\
&\mathbb{I}_{3,3}^{-1}(\boldsymbol{\Theta}) = \frac{24\sigma_Y^2}{Y_m^2 N(N^2-1)}\left(\frac{f_s}{2\pi}\right)^2.
\end{aligned}
\tag{5.16}
$$

Finally, the CRBs for the variances of an estimator of Y_m, θ, and f are given by $\mathbb{I}_{1,1}^{-1}(\boldsymbol{\Theta})$, $\mathbb{I}_{2,2}^{-1}(\boldsymbol{\Theta})$, and $\mathbb{I}_{3,3}^{-1}(\boldsymbol{\Theta})$ in (5.16), respectively, after substituting Y_m with its estimate $\hat{Y}_m$, and can be summarized as follows:

$$
\begin{aligned}
&\mathrm{CRB}(\hat{Y}_m) = \frac{2\sigma_Y^2}{N}; \quad \mathrm{CRB}(\hat{\theta}) = \frac{4\sigma_Y^2(2N+1)}{\hat{Y}_m^2 N(N-1)},\\
&\mathrm{CRB}(\hat{f}) = \left(\frac{f_s}{2\pi}\right)^2 \frac{24\sigma_Y^2}{\hat{Y}_m^2 N(N^2-1)}.
\end{aligned}
\tag{5.17}
$$

5.2.2 Expressions for variance of the parameter estimates

Following the statistical analysis given in [108], the variances of $\hat{Y}_m$, $\hat{\theta}$, and $\hat{f}$ are found to be approximately two, six, and two times the CRBs derived in (5.17), respectively; and hence, the estimated variances of $\hat{f}$, $\hat{\theta}$, and $\hat{Y}_m$ are given by $\hat{\sigma}_f^2$ (in p.u.), $\hat{\sigma}_{Y_m}^2$ (in p.u.), and $\hat{\sigma}_\theta^2$ (in rad^2), respectively, as follows:

$$
\hat{\sigma}_f^2 = \frac{2\mathrm{CRB}(\hat{f})}{f_0^2}, \quad \hat{\sigma}_{Y_m}^2 = 2\mathrm{CRB}(\hat{Y}_m), \quad \hat{\sigma}_\theta^2 = 6\mathrm{CRB}(\hat{\theta}), \tag{5.18}
$$

where $\mathrm{CRB}(\hat{f})$, $\mathrm{CRB}(\hat{Y}_m)$, and $\mathrm{CRB}(\hat{\theta})$ are given by (5.17).

5.3 IMPLEMENTATION EXAMPLE

The 16-machine, 68-bus test system (Fig. A.1) described in Appendix A has been used as an implementation example for demonstrating the estimator developed in this chapter, with modeling and simulation being performed on MATLAB-Simulink using ode45 solver.

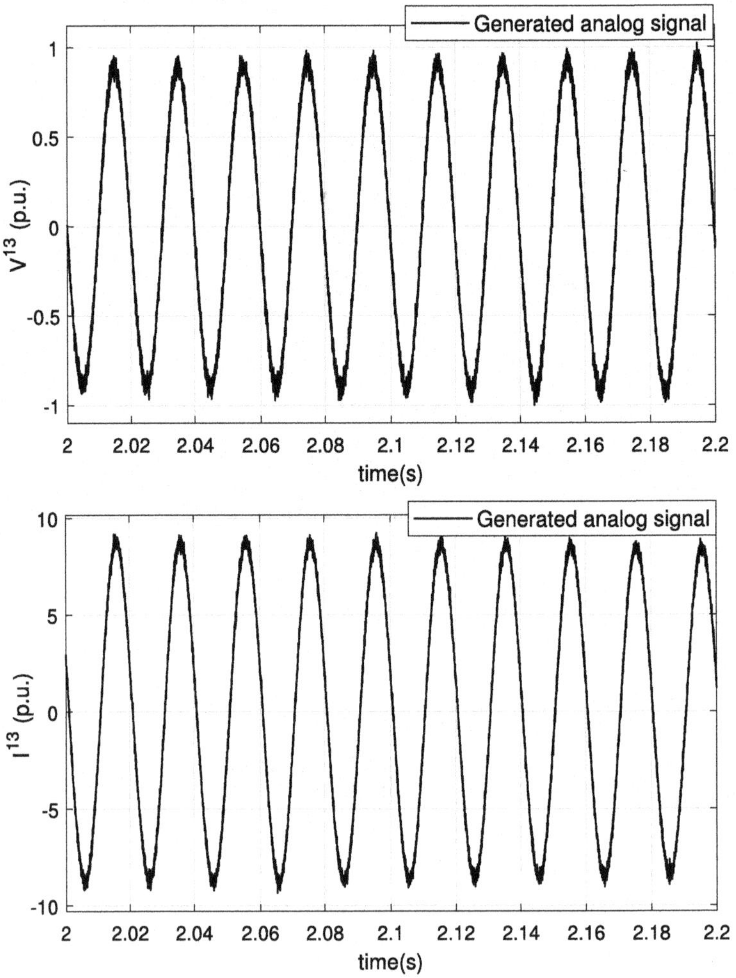

Figure 5.1 Generated analogue signals of current and voltage for the 13th unit.

In the simulation, the interpolated DFT-based signal parameter estimator (developed in Sections 5.1–5.2) runs at each generation unit. The measurements which are required by the estimator are $V(t)$ and $I(t)$, and they are generated by adding noise to the simulated analogue values of terminal voltage and current of the unit. Since the IEEE [102] and IEC [101] standards specify that the measurement error in CTs/VTs should be less than 3%, in order to simulate the worst-case scenario in terms of noise, the standard deviation of noise in measurements of $V(t)$ and $I(t)$

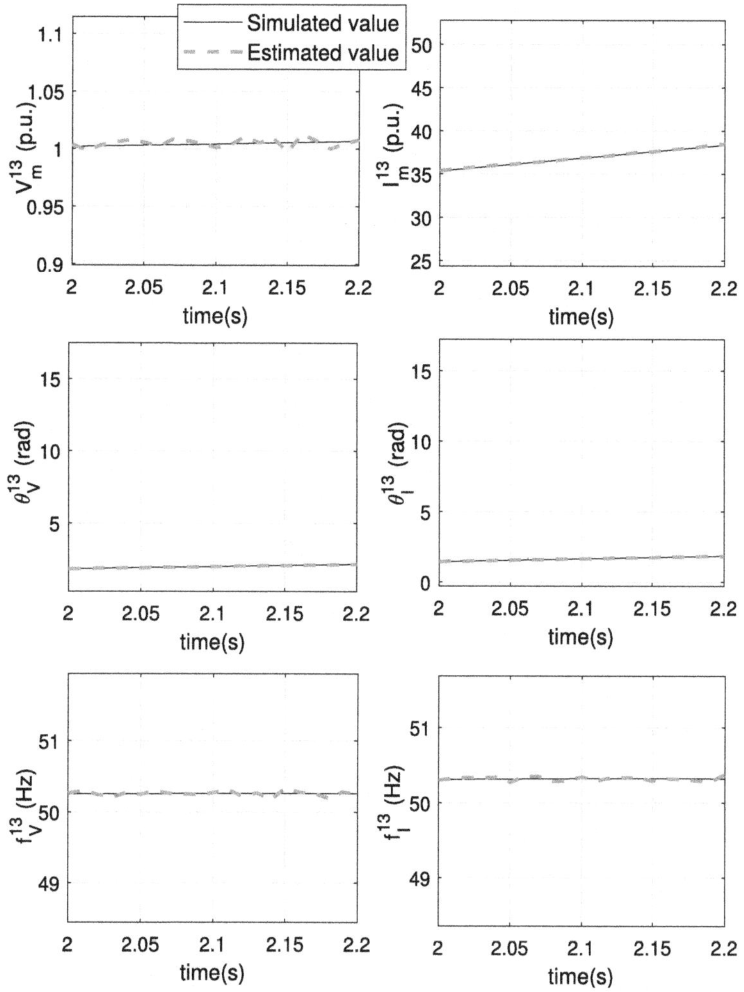

Figure 5.2 Simulated vs. estimated signal parameters of V and I.

is taken as 3% of the amplitudes of these signals. Also, as explained in Section 5.1, N, f_0, and f_s are taken as 1200, 50 Hz, and 40000 Hz, respectively.

The system starts from a steady state in the simulation. Then at $t = 1$ s, a disturbance is created by a three-phase fault at bus 54 and is cleared after 0.18 s by opening one of the tie lines between buses 53–54. The system is simulated for 15 s. The simulated measurements of $V(t)$ and $I(t)$ are shown in Fig. 5.1 for a small time duration (from 2 s to 2.2 s) for the

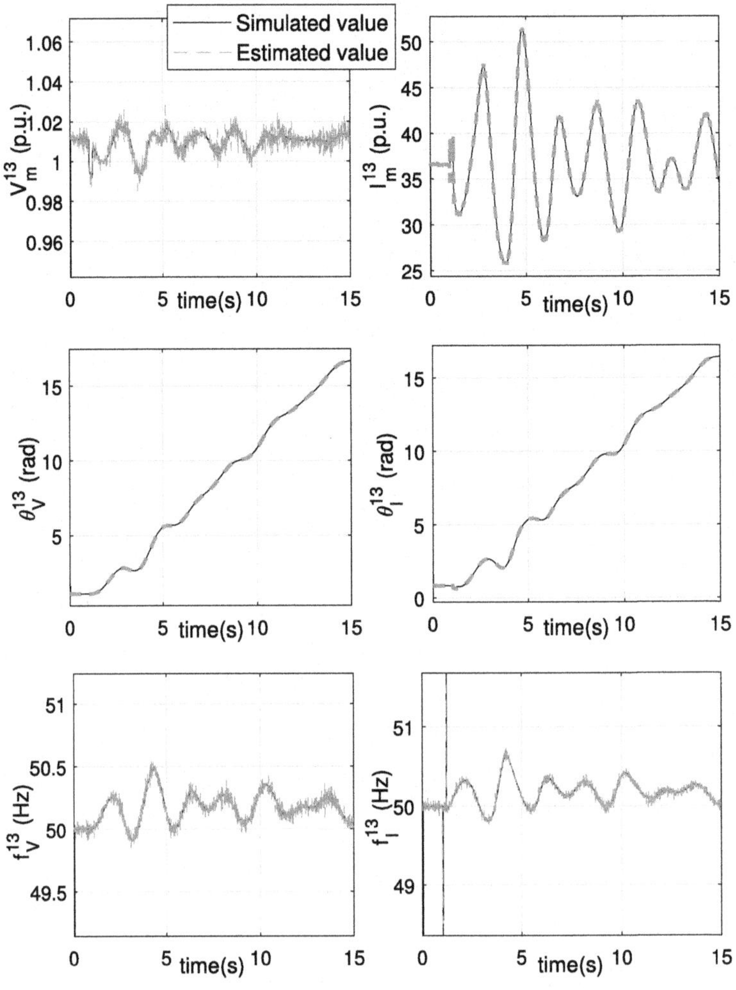

Figure 5.3 Simulated vs. estimated signal parameters of V and I for the complete duration of the simulation.

13th unit. The estimates $\hat{V}_m$, $\hat{f}_V$, and $\hat{\theta}_V$ for $V(t)$, and $\hat{I}_m$, $\hat{f}_I$, and $\hat{\theta}_I$ for $I(t)$ obtained using the interpolated DFT estimator are shown in Fig. 5.2 for the same duration of 2 s to 2.2 s, while for the complete duration of the simulation, the estimates are plotted in Fig. 5.3. Corresponding estimation errors, which are the differences between the estimated and simulated values, are plotted in Fig. 5.4 for the complete duration of the simulation.

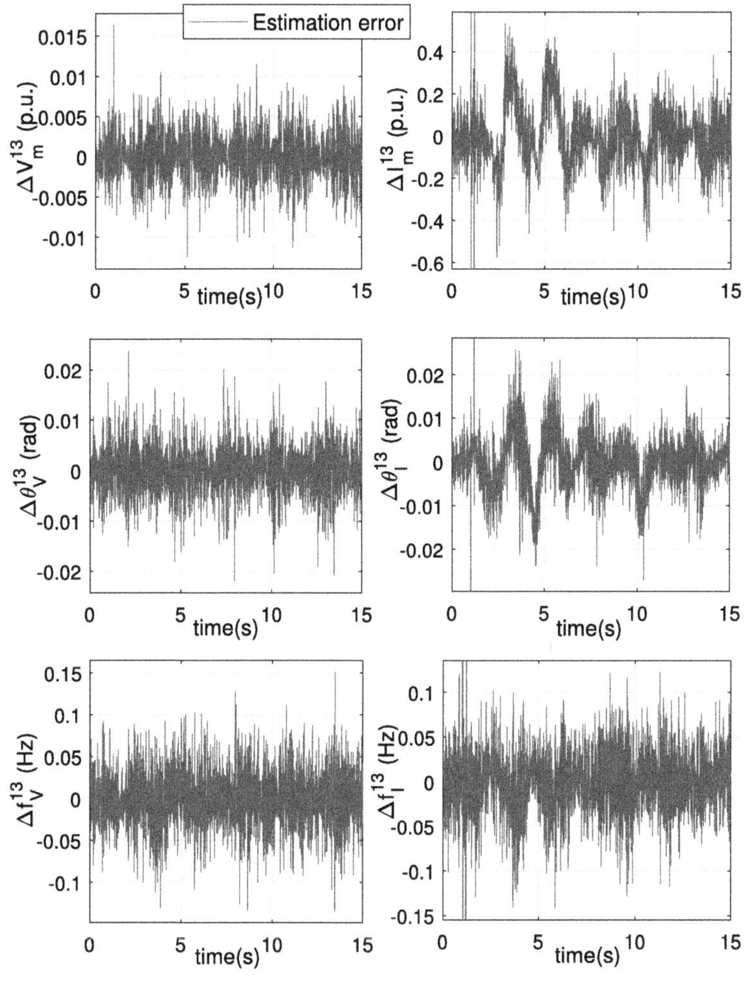

Figure 5.4 Estimation errors in signal parameter estimates of V and I for the complete duration of the simulation.

It can be observed in Fig. 5.2 and Fig. 5.3 that the signal parameter estimates almost coincide with the simulated values; hence, interpolated DFT is an accurate and stable method for real-time signal parameter estimation. Also, because of larger fluctuations in I_m as compared to V_m, the estimates for V_m appear to be much noisier than those for I_m, but actually the percentage noise in both remains almost the same, as can be observed in Fig. 5.4. Moreover, the larger fluctuations in I_m lead to larger fluctuations in noises in all the three estimates for $I(t)$, as can be seen in Fig. 5.4.

5.4 SUMMARY

This chapter presents an interpolated DFT-based algorithm for estimating the parameters of analogue voltages and currents which are measured at a bus in a power system using CT/VTs, the parameters being the means and variances of magnitude, phase, and frequency of the sinusoidal voltage or current signal. A detailed derivation of the mathematical expressions for all these parameters has also been presented. The estimation method has been demonstrated for one of the generation units of the 68-bus test system.

CHAPTER 6

Decentralized Dynamic Estimation Using CTs/VTs

It was explained in Chapter 4 that in order to monitor and control the oscillations and related dynamics which cause instability in power system, the operating state of the system needs to be estimated in real-time, with update rates which are in time scales of ten milliseconds or less (as the time constants associated with such oscillations are not more than ten milliseconds), and this real-time estimation of the operating state is known as dynamic state estimation (DSE). DSE is a fast growing and widely researched field [25–41], and it lays the foundation for a new generation of control methods which can dynamically stabilize the system.

The dynamic states which are estimated and obtained as outputs from DSE algorithms are angles, speeds, voltages, and fluxes of the rotors of all the generators in the power system. The inputs which are given to DSE algorithms are some time-varying quantities, such as voltage and current of the stator, and some time-invariant quantities, such as resistance, reactance, inertia, and other constants of the generator. The constant quantities are measured or estimated beforehand and are used as parameters in DSE algorithms.

A generator's voltage, current, and power are sinusoidal quantities, and since each sinusoid has a magnitude and a phase (which are together known as a phasor), these quantities can either be represented as sine waves or as phasors. The conversion of sine waves to phasors is done by phasor measurement units (PMUs). During this conversion, PMUs provide a common reference angle to the phase of the phasor. This is required because a power system is a rotational system (because of the rotational parts of the generators), and every rotational system needs to have a reference angle which is common for all the angles in the system. This common reference angle is provided by PMUs using a common time source for synchronization which is obtained using the time clock of the global positioning system (GPS) [112], as mentioned in Chapter 4. An important dynamic state which also requires this common reference angle is the rotor angle of a machine. Thus, in order to estimate the rotor angle, any PMU-based DSE algorithm

Dynamic Estimation and Control of Power Systems
https://doi.org/10.1016/B978-0-12-814005-5.00017-0

requires time-synchronized measurements [25–41], and the decentralized DSE algorithm presented in Chapter 4 is one such algorithm.

A problem with time synchronization is that it has associated noise and synchronization errors [112]. Synchronization errors increase the total vector error (TVE) of PMU measurements. For instance, in [113] it was demonstrated that a time synchronization error of just 10 μs, well within the maximum allowable IEEE standard of 31.6 μs, can make the TVE greater than 1% for PMUs, which is the IEEE standard limit for PMUs [103,104]. As synchronized measurements are used for DSE, these errors can get propagated to the estimated states and deteriorate the overall accuracy and robustness of estimation. It is also not possible to completely eliminate time synchronization as it is inherently required for estimation of rotor angles. This leads to the main idea of this chapter: *although time synchronization is needed for estimation of rotor angle, it is not needed for estimation of other dynamic states, such as rotor speed, rotor voltages, and fluxes, as these states are not defined with respect to a common reference angle. Thus, if the dynamic model which is used for estimation can be modified in such a way that rotor angle is replaced with another angle which does not require time synchronization, then this can minimize the effects of synchronization on accuracy and robustness of estimation.*

The method presented in this chapter provides an algorithm for DSE which realizes the above idea. The rest of the chapter is organized as follows. Section 6.1 specifies the decoupled equations which are used in the developed method. Section 6.2 explains how the parameter estimates obtained in Chapter 5 can be further used for DSE using unscented Kalman filtering. Section 6.3 presents simulations to demonstrate the developed estimation method. Although Sections 6.1–6.3 consider a balanced three-phase system, Section 6.4 very briefly describes how this method can be extended to a general unbalanced three-phase system. Section 6.5 summarizes the chapter.

6.1 DECOUPLED POWER SYSTEM EQUATIONS AFTER INCORPORATING INTERNAL ANGLE

In order to not depend on the time synchronization from DSE, the estimation model is modified to incorporate a relative angle (which does not require synchronization) instead of a rotor angle. One such angle is the difference between the rotor angle and the generator terminal voltage phase, also known as the internal angle of the generator. As the rotor angle and the voltage phase have a common reference angle, this reference angle gets canceled in the difference of the two quantities. Thus, the internal angle, rotor speed, voltages, and fluxes can be estimated using the modified estimation model without requiring any synchronized measurements. These dynamic states can then be utilized for decentralized control of the generator (as presented in Chapters 7–9). It should be noted that if the estimation of rotor angle is specifically required then it can be indirectly estimated as the sum of the estimated internal angle and the measured terminal voltage phase obtained using PMU.

In order to estimate the internal angle instead of estimating the rotor angle, the decoupled equations and the pseudoinputs for a generator also get altered. The altered decoupled equations are given by (6.1)–(6.11), derived using the subtransient model of machines described in Chapter 3. In these equations, the altered pseudoinputs are V_m and voltage frequency, f_V. Slow dynamics of the speed governor have been ignored, although they can be added if required. Also, a model of a static IEEE-ST1A excitation system is included with the model of each machine. We have

$$\Delta\dot{\alpha}^i = (\omega^i - f_V^i), \tag{6.1}$$

$$\Delta\dot{\omega}^i = \frac{\omega_0}{2H^i}(T_{m0}^i - T_e^i) - \frac{D^i}{2H^i}\Delta\omega^i, \tag{6.2}$$

$$\dot{E}_d^{\prime i} = \frac{1}{T_{q0}^{\prime i}}[-E_d^{\prime i} - (X_q^i - X_q^{\prime i})[K_{q1}^i I_q^i + K_{q2}^i \frac{\Psi_{2q}^i + E_d^{\prime i}}{X_q^{\prime i} - X_l^i}]], \tag{6.3}$$

$$\dot{E}_q^{\prime i} = \frac{E_{fd}^i - E_q^{\prime i} + (X_d^i - X_d^{\prime i})[K_{d1}^i I_d^i + K_{d2}^i \frac{\Psi_{1d}^i - E_q^{\prime i}}{X_d^{\prime i} - X_l^i}]}{T_{d0}^{\prime i}}, \tag{6.4}$$

$$\dot{\Psi}_{1d}^i = \frac{1}{T_{d0}^{\prime\prime i}}[E_q^{\prime i} + (X_d^{\prime i} - X_l^i)I_d^i - \Psi_{1d}^i], \tag{6.5}$$

$$\dot{\Psi}_{2q}^i = \frac{1}{T_{q0}^{\prime\prime i}}[-E_d^{\prime i} + (X_q^{\prime i} - X_l^i)I_q^i - \Psi_{2q}^i], \tag{6.6}$$

$$\dot{V}_r^i = \frac{1}{T_r^i}[V_m^i - V_r^i], \text{ where} \tag{6.7}$$

$$E_{fd}^i = K_a^i[V_{ref}^i - V_r^i], \; E_{fdmin}^i \le E_{fd}^i \le E_{fdmax}^i, \tag{6.8}$$

$$\begin{bmatrix} I_d^i \\ I_q^i \end{bmatrix} = \begin{bmatrix} R_s^i & X_q^{\prime\prime i} \\ -X_d^{\prime\prime i} & R_s^i \end{bmatrix}^{-1} \begin{bmatrix} E_d^{\prime i}K_{q1}^i - \Psi_{2q}^i K_{q2}^i - V_d^i \\ E_q^{\prime i}K_{d1}^i + \Psi_{1d}^i K_{d2}^i - V_q^i \end{bmatrix}, \tag{6.9}$$

$$T_e^i = \frac{\omega_0}{\omega^i} P_e^i, \; P_e^i = V_d^i I_d^i + V_q^i I_q^i = V_m^i I_m^i \cos(\theta_V^i - \theta_I^i), \tag{6.10}$$

$$I_m^i = \sqrt{I_d^{i2} + I_q^{i2}}, \; V_d^i = -V_m^i \sin\alpha^i, \; V_q^i = V_m^i \cos\alpha^i. \tag{6.11}$$

The above equations can be written in the following composite state space form, which will be used for the developed robust DSE (here pseudoinputs are denoted by $\boldsymbol{u}'^i$, and the process noise and the noise in pseudoinputs have been included and denoted by $\boldsymbol{v}^i$ and $\boldsymbol{w}'^i$, respectively):

$$\begin{aligned} \dot{\boldsymbol{x}}^i &= \boldsymbol{g}'^i(\boldsymbol{x}^i, \boldsymbol{u}'^i, \boldsymbol{w}'^i) + \boldsymbol{v}^i; \; \boldsymbol{u}'^i - \boldsymbol{w}'^i = [V_m^i \; f_V^i]^T, \\ \boldsymbol{x}^i &= [\alpha^i \; \omega^i \; E_d'^i \; E_q'^i \; \Psi_{1d}^i \; \Psi_{2q}^i \; V_r^i]^T. \end{aligned} \tag{6.12}$$

6.2 TWO-STAGE ESTIMATION BASED ON INTERPOLATED DFT AND UKF

The estimation is divided into two stages. In the first stage, estimates of means and variances of the parameters of the generator's terminal voltage and current are obtained using the interpolated-DFT method described in Chapter 5. These estimates are given as inputs to the second stage, which is the UKF stage. The advantage of obtaining the analytical expressions for the mean values of the parameters ($\hat{f}$, $\hat{\theta}$, and $\hat{Y}_m$) and their variances ($\hat{\sigma}_f^2$, $\hat{\sigma}_{Y_m}^2$, and $\hat{\sigma}_\theta^2$) in (5.11)–(5.18) in Chapter 5 is that these means and variances can be continuously updated and provided to the UKF stage, thereby improving the accuracy of the dynamic state estimation.

As explained in Chapter 4, UKF is a discrete method and, hence, the system given by (6.12) needs to be discretized before UKF can be applied to it. Discretizing (6.12) at a sampling period T_0, by approximating $\dot{\boldsymbol{x}}^i$ with $(\boldsymbol{x}^{ik} - \boldsymbol{x}^{i\bar{k}})/T_0$, gives the following equation (where k and $\bar{k}$ represent the kth and $(k-1)$th samples, respectively):

$$\begin{aligned} \boldsymbol{x}^{ik} &= \boldsymbol{x}^{i\bar{k}} + T_0 \boldsymbol{g}'^i(\boldsymbol{x}^{i\bar{k}}, \boldsymbol{u}'^{i\bar{k}}, \boldsymbol{w}'^{i\bar{k}}) + \boldsymbol{v}^{i\bar{k}} \\ \Rightarrow \boldsymbol{x}^{ik} &= \boldsymbol{g}^i(\boldsymbol{x}^{i\bar{k}}, \boldsymbol{u}'^{i\bar{k}}, \boldsymbol{w}'^{i\bar{k}}) + \boldsymbol{v}^{i\bar{k}}. \end{aligned} \tag{6.13}$$

In the above model, $\hat{V}_m^{ik}$ and $\hat{f}_V^{ik}$ (found using the interpolated-DFT method) are used in the pseudoinput vector $\boldsymbol{u}'^{ik}$ as follows:

$$\boldsymbol{u}'^{ik} = [\hat{V}_m^{ik} \; \hat{f}_V^{ik}]^T = [V_m^{ik} \; f_V^{ik}]^T + \boldsymbol{w}'^{ik}. \tag{6.14}$$

UKF also requires a measurement model besides the above process model. The estimates of active power, P_e^{ik} (defined by (6.10)), and stator current magnitude, I_m^{ik} (defined by (6.11)), which are obtained using the interpolated-DFT method are used as measurements for UKF. After incorporating the measurement noise, $\boldsymbol{w}^{ik}$, the measurement model is given as follows:

$$\boldsymbol{y}^{ik}=\begin{bmatrix}\hat{P}_e^{ik}\\ \hat{I}_m^{ik}\end{bmatrix}=\begin{bmatrix}V_d^{ik}I_d^{ik}+V_q^{ik}I_q^{ik}\\ \sqrt{I_d^{ik^2}+I_q^{ik^2}}\end{bmatrix}+\boldsymbol{w}^{ik}$$
$$\Rightarrow \boldsymbol{y}^{ik}=\boldsymbol{h}^i(\boldsymbol{x}^{ik},\boldsymbol{u}'^{ik},\boldsymbol{w}'^{ik})+\boldsymbol{w}^{ik}, \tag{6.15}$$

where $\boldsymbol{u}'^{ik}$ and $\boldsymbol{y}^{ik}$ are estimated quantities and have finite variances which need to be included in the process and measurement models, respectively. This is done by including $\boldsymbol{w}'^{ik}$ and $\boldsymbol{w}^{ik}$ in the models as the following zero-mean noises:

$$\boldsymbol{w}'^{ik}=\begin{bmatrix}w'_{V_m^{ik}}\\ w'_{f_V^{ik}}\end{bmatrix},\ \hat{\boldsymbol{w}}'^{ik}=\begin{bmatrix}0\\0\end{bmatrix},\ \boldsymbol{P}_{w'}^{ik}=\begin{bmatrix}\hat{\sigma}^2_{V_m^{ik}} & 0\\ 0 & \hat{\sigma}^2_{f_V^{ik}}\end{bmatrix}, \tag{6.16}$$

$$\boldsymbol{w}^{ik}=\begin{bmatrix}w_{P_e^{ik}}\\ w_{I_m^{ik}}\end{bmatrix},\ \hat{\boldsymbol{w}}^{ik}=\begin{bmatrix}0\\0\end{bmatrix},\ \boldsymbol{P}_{w}^{ik}=\begin{bmatrix}\hat{\sigma}^2_{P_e^{ik}} & 0\\ 0 & \hat{\sigma}^2_{I_m^{ik}}\end{bmatrix}, \tag{6.17}$$

where $\boldsymbol{P}_{w'}^{ik}$ and $\boldsymbol{P}_{w}^{ik}$ denote the covariance matrices of $\boldsymbol{w}'^{ik}$ and $\boldsymbol{w}^{ik}$, respectively. In order to find the estimates and variances in (6.14)–(6.17), the stator voltage, $V^i(t)$, and stator current, $I^i(t)$, measured using VT and CT, respectively, are processed using the DFT method. Thus, $\hat{V}_m^{ik}$, $\hat{f}_V^{ik}$, $\hat{\theta}_V^{ik}$, $\hat{\sigma}^2_{V_m^{ik}}$, $\hat{\sigma}^2_{f_V^{ik}}$, and $\hat{\sigma}^2_{\theta_V^{ik}}$ are obtained by putting $Y(t)=V^i(t)$ in (5.1)–(5.10) and updating these estimates and variances for every kth sample. Similarly, $\hat{I}_m^{ik}$, $\hat{f}_I^{ik}$, $\hat{\theta}_I^{ik}$, $\hat{\sigma}^2_{I_m^{ik}}$, $\hat{\sigma}^2_{f_I^{ik}}$, and $\hat{\sigma}^2_{\theta_I^{ik}}$ are obtained by putting $Y(t)=I^i(t)$. As $P_e^{ik}=V_m^{ik}I_m^{ik}\cos(\theta_V^{ik}-\theta_I^{ik})$ (from (6.10)) and the mean values and variances of V_m^{ik}, I_m^{ik}, θ_V^{ik}, and θ_I^{ik} are known, the mean value of P_e^{ik} (denoted as $\hat{P}_e^{ik}$) and its estimated variance (denoted as $\hat{\sigma}^2_{P_e^{ik}}$) can be represented in terms of these known quantities and have been obtained as follows (here it should be noted that by definition θ_V^{ik} and θ_I^{ik} lie in the interval $(-\pi/2,\pi/2]$, hence, they should be "unwrapped" by adding or subtracting suitable multiples of

π to them, in order to find $\cos(\theta_V^{ik} - \theta_I^{ik})$):

$$\hat{P}_e^{ik} = \hat{V}_m^{ik} \hat{I}_m^{ik} \cos(\hat{\theta}_V^{ik} - \hat{\theta}_I^{ik}),$$

$$\hat{\sigma}^2_{P_e^{ik}} = [\hat{\sigma}^2_{V_m^{ik}} (\hat{I}_m^{ik})^2 + (\hat{V}_m^{ik})^2 \hat{\sigma}^2_{I_m^{ik}}] \cos^2(\hat{\theta}_V^{ik} - \hat{\theta}_I^{ik}) \qquad (6.18)$$

$$+ (\hat{V}_m^{ik})^2 (\hat{I}_m^{ik})^2 [\hat{\sigma}^2_{\theta_V^{ik}} + \hat{\sigma}^2_{\theta_I^{ik}}] \sin^2(\hat{\theta}_V^{ik} - \hat{\theta}_I^{ik}).$$

Thus, the four quantities which are required by the UKF stage from the DFT stage are $\boldsymbol{u}'^i$, $\boldsymbol{y}^i$, $\boldsymbol{P}_{w'}^{ik}$, and $\boldsymbol{P}_w^{ik}$, given by (6.14)–(6.17). These quantities should be updated every T_0 s, as this is the sampling period of the UKF stage. Also, in (6.13), both $\boldsymbol{x}^{ik}$ and $\boldsymbol{w}'^{ik}$ are unknown quantities and can be combined as a composite state vector $\boldsymbol{X}^{ik}$ with a composite covariance matrix $\boldsymbol{P}_{\boldsymbol{X}}^{ik}$ defined as follows:

$$\boldsymbol{X}^{ik} = \begin{bmatrix} \boldsymbol{x}^{ik} \\ \boldsymbol{w}'^{ik} \end{bmatrix}, \ \hat{\boldsymbol{X}}^{ik} = \begin{bmatrix} \hat{\boldsymbol{x}}^{ik} \\ \hat{\boldsymbol{w}}'^{ik} \end{bmatrix}, \ \boldsymbol{P}_{\boldsymbol{X}}^{ik} = \begin{bmatrix} \boldsymbol{P}_{\boldsymbol{x}}^{ik} & \boldsymbol{P}_{\boldsymbol{x}\boldsymbol{w}'}^{ik} \\ \boldsymbol{P}_{\boldsymbol{x}\boldsymbol{w}'}^{ik\,T} & \boldsymbol{P}_{\boldsymbol{w}'}^{ik} \end{bmatrix}. \qquad (6.19)$$

Here $\boldsymbol{P}_{\boldsymbol{x}}^{ik}$ is the covariance matrix of $\boldsymbol{x}^{ik}$ and $\boldsymbol{P}_{\boldsymbol{x}\boldsymbol{w}'}^{ik}$ is the crosscovariance matrix of $\boldsymbol{x}^{ik}$ and $\boldsymbol{w}'^{ik}$. With the above definition, the model in (6.13)–(6.15) is redefined as follows:

$$\boldsymbol{X}^{ik} = \boldsymbol{g}^i(\boldsymbol{X}^{i\bar{k}}, \boldsymbol{u}'^{i\bar{k}}) + \boldsymbol{v}^{i\bar{k}},$$
$$\boldsymbol{y}^{ik} = \boldsymbol{h}^i(\boldsymbol{X}^{ik}, \boldsymbol{u}'^{ik}) + \boldsymbol{w}^{ik}. \qquad (6.20)$$

With (6.20) as model and $\boldsymbol{x}^{i0}$ as steady-state estimate of $\boldsymbol{x}^{ik}$ and with the knowledge of $\boldsymbol{g}^i$, $\boldsymbol{h}^i$, $\boldsymbol{u}'^i$, $\boldsymbol{y}^i$, $\boldsymbol{P}_{w'}^{ik}$, $\boldsymbol{P}_w^{ik}$, and the process noise covariance matrix, $\boldsymbol{P}_v^{ik}$, the filtering equations of UKF for the kth iteration and ith unit are given as follows.

STEP 1: Initialize

if (k==1) **then** initialize $\hat{\boldsymbol{x}}^{i\bar{k}} = \boldsymbol{x}^{i0}$, $\hat{\boldsymbol{w}}'^{i\bar{k}} = \boldsymbol{0}_{2\times1}$, $\boldsymbol{P}_{\boldsymbol{x}}^{i\bar{k}} = \boldsymbol{P}_v^{i0}$, $\boldsymbol{P}_{\boldsymbol{x}\boldsymbol{w}}'^{i\bar{k}} = \boldsymbol{0}_{m^i\times2}$, $\boldsymbol{P}_w'^{i\bar{k}} = \boldsymbol{P}_w'^{i0}$ in (6.19) to get $\boldsymbol{P}_{\boldsymbol{X}}^{i\bar{k}}$ & $\hat{\boldsymbol{X}}^{i\bar{k}}$.

else reinitialize $\hat{\boldsymbol{w}}'^{i\bar{k}}$ and $\boldsymbol{P}_{w'}^{i\bar{k}}$ in (6.19) according to (6.16), leaving the rest of the elements in $\hat{\boldsymbol{X}}^{i\bar{k}}$ and $\boldsymbol{P}_{\boldsymbol{X}}^{i\bar{k}}$ unchanged.

STEP 2: Generate sigma points

$$\chi_l^{i\bar{k}} = \hat{\boldsymbol{X}}^{i\bar{k}} + \left(\sqrt{n^i \boldsymbol{P}_{\boldsymbol{X}}^{i\bar{k}}}\right)_l, \quad l = 1, 2, \ldots, n^i,$$

$$\chi_l^{i\bar{k}} = \hat{X}^{i\bar{k}} - \left(\sqrt{n^i P_X^{i\bar{k}}}\right)_l, \quad l = (n^i+1), (n^i+2), \ldots, 2n^i$$

STEP 3: Predict states

$$\chi_l^{ik^-} = g^i(\chi_l^{i\bar{k}}, u'^{i\bar{k}}); \ \hat{X}^{ik^-} = \frac{1}{2n^i}\sum_{l=1}^{2n^i} \chi_l^{ik^-}$$

$$P_X^{ik^-} = \frac{1}{2n_i}\sum_{l=1}^{2n^i}[\chi_l^{ik^-} - \hat{X}^{ik^-}][\chi_l^{ik^-} - \hat{X}^{ik^-}]^T + P_v^{ik}$$

STEP 4: Predict measurements

$$y_l^{ik^-} = h^i(\chi_l^{i\bar{k}}, u'^{i\bar{k}}); \ \hat{y}^{ik^-} = \frac{1}{2n^i}\sum_{l=1}^{2n^i} y_l^{ik^-}$$

$$P_y^{ik^-} = \frac{1}{2n_i}\sum_{l=1}^{2n^i}[y_l^{ik^-} - \hat{y}^{ik^-}][y_l^{ik^-} - \hat{y}^{ik^-}]^T + P_w^{ik}$$

$$P_{Xy}^{ik^-} = \frac{1}{2n_i}\sum_{l=1}^{2n^i}[\chi_l^{ik^-} - \hat{X}^{ik^-}][y_l^{ik^-} - \hat{y}^{ik^-}]^T$$

STEP 5: Kalman update

$$K^{ik} = P_{Xy}^{ik^-}(P_y^{ik^-})^{-1}; \quad \hat{X}^{ik} = \hat{X}^{ik^-} + K^{ik}(y^{ik} - \hat{y}^{ik^-})$$

$$P_X^{ik} = P_X^{i\bar{k}} - K^{ik}[P_{Xy}^{ik^-}]^T$$

STEP 6: Output and time update

output $\hat{X}^{ik}$ and P_X^{ik}, $k \leftarrow (k+1)$, goto STEP 1.

6.3 CASE STUDY

The 16-machine, 68-bus test system (Fig. A.1) described in Appendix A has been used for the case study and MATLAB-Simulink (using ode45 solver) has been used for its modeling and simulation.

6.3.1 Simulation parameters

The robust dynamic state estimator (developed in Section 6.1 and Section 6.2) runs at the location of each generation unit and provides dynamic state estimates for the unit. The measurements which are required by the estimator are $V(t)$ and $I(t)$ and they are generated by adding noise to the simulated analogue values of terminal voltage and current of the unit. As explained in Section 5.1, N, f_0, and f_s are taken as 1200, 50 Hz, and 40000 Hz, respectively. The sampling period of the UKF stage, T_0, is taken as 0.01 s, as explained in Chapter 4; and thus, the estimates obtained from the DFT stage are also updated every 0.01 s. Also, P_v^{ik} is found as described in Chapter 4. For comparison with the developed estimator, another UKF-based dynamic state estimator which uses PMU measurements (as presented in Chapter 4) also runs at each unit's location and is termed DSE-with-PMU method. An estimate of the internal angle in the case of DSE-with-PMU is obtained by subtracting the measurement of the terminal voltage phase from the estimate of the rotor angle.

The measurement error for the robust DSE method is the percentage error in the analogue signals of $V(t)$ and $I(t)$, while the measurement error for the DSE-with-PMU method is the TVE in the phasor measurements of the terminal voltage and current. As the measurement errors for the two estimators are of two different kinds, these methods cannot be directly compared for the same noise levels. Nevertheless, the performance of the two methods for standard measurement errors can be compared, as specified by IEEE [103,104], [102] and IEC [101]. As mentioned in these standards, the measurement error in CTs/VTs should be less than 3%, while the standard error for PMUs is 1% TVE. Hence, in the base case for comparison, the measurement error for robust DSE is taken as 3%, while for the DSE-with-PMU method, it is taken as 1% TVE.

The system starts from a steady state in the simulation. Then at $t = 1$ s, a disturbance is created by a three-phase fault at bus 54 and is cleared after 0.18 s by opening of one of the tie lines between buses 53–54. The simulated states, along with their estimated values for the base case for one of the units (the 13th unit), are plotted in Fig. 6.1 and Fig. 6.3. The corresponding estimation errors are plotted in Fig. 6.2 and Fig. 6.4.

6.3.2 Estimation accuracy

It can be seen in Figs. 6.1–6.4 that for robust DSE the plots of estimated values almost coincide with those of the simulated values and the estimation errors are low, but for the DSE-with-PMU method the difference between the simulated and estimated values is apparent and the estimation errors are much higher. This shows that the robust DSE method performs accurately with standard measurement errors in CTs/VTs, while DSE-with-PMU fails to do so with standard errors in PMU measurements.

Robustness of the robust DSE method has been tested against varying noise levels in measurements. Fig. 6.5 shows the estimation results for ω^{13} for two more cases: in the first case the noise levels are one-third the base case, while in the second case the noise levels are thrice the base case. Also, root mean-squared errors (RMSEs) for varying error levels have been calculated and tabulated in Table 6.1 and Table 6.2 for the two methods. It can be observed that the performance of the robust DSE method remains robust to errors up to 3%, and even for an error of 10%, its performance deteriorates only to a small extent. On the other hand, the DSE-with-PMU

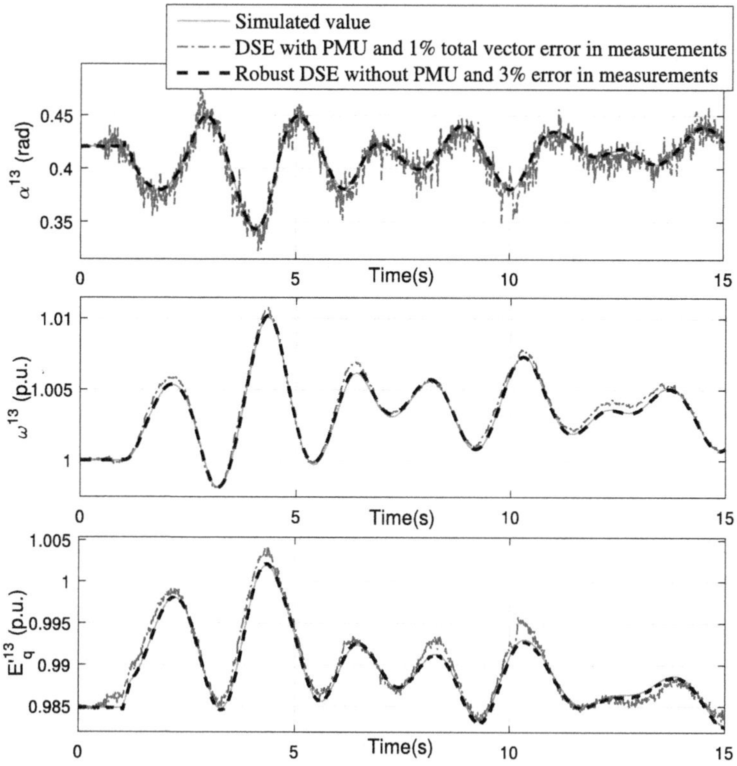

Figure 6.1 Comparison of DSE for α^{13}, ω^{13}, and $E_q'^{13}$ for the base case.

Table 6.1 Root mean-square errors for robust DSE

Estimation of RMSEs for various measurement errors (MEs) (in p.u.)

State	0.3% ME	1% ME	3% ME	10% ME
α^{13}	4.85×10^{-4}	4.97×10^{-4}	5.80×10^{-4}	4.94×10^{-3}
ω^{13}	3.42×10^{-5}	5.66×10^{-5}	7.81×10^{-5}	3.97×10^{-4}
$E_q'^{13}$	1.44×10^{-4}	1.50×10^{-4}	1.76×10^{-4}	3.15×10^{-3}
$E_d'^{13}$	3.22×10^{-4}	3.27×10^{-4}	3.31×10^{-4}	4.13×10^{-3}
ψ_{2q}^{13}	3.54×10^{-4}	3.67×10^{-4}	4.18×10^{-4}	4.51×10^{-3}
ψ_{1d}^{13}	1.92×10^{-4}	2.25×10^{-4}	4.25×10^{-4}	2.63×10^{-3}
V_r^{13}	6.92×10^{-4}	1.56×10^{-3}	4.54×10^{-3}	1.38×10^{-2}

method does not perform accurately for error levels above 0.3% TVE, that is, it is not accurate for 1% TVE and 3% TVE.

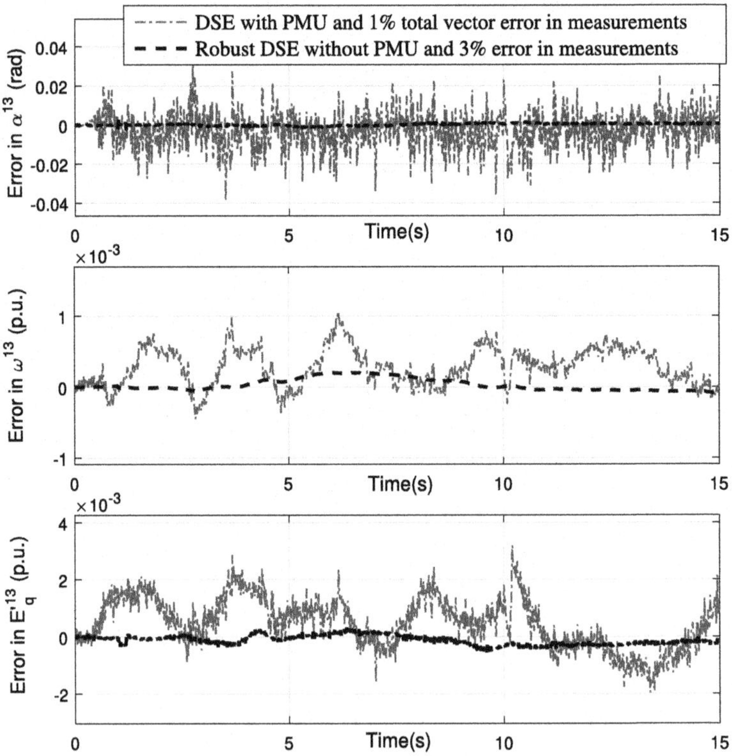

Figure 6.2 Estimation errors for α^{13}, ω^{13}, and $E_q'^{13}$ for the base case.

Table 6.2 Root mean-square errors for DSE-with-PMU

Estimation of RMSEs for various measurement TVEs (in p.u.)				
State	**0.1% TVE**	**0.3% TVE**	**1% TVE**	**3% TVE**
α^{13}	8.06×10^{-4}	2.31×10^{-3}	7.19×10^{-3}	3.70×10^{-2}
ω^{13}	7.04×10^{-5}	9.44×10^{-5}	4.14×10^{-4}	1.95×10^{-3}
$E_q'^{13}$	1.22×10^{-4}	3.58×10^{-4}	1.23×10^{-3}	6.49×10^{-3}
$E_d'^{13}$	5.13×10^{-4}	7.11×10^{-4}	1.87×10^{-3}	8.95×10^{-3}
ψ_{2q}^{13}	6.31×10^{-4}	8.82×10^{-4}	2.81×10^{-3}	1.39×10^{-2}
ψ_{1d}^{13}	2.36×10^{-4}	6.31×10^{-4}	2.21×10^{-3}	1.34×10^{-2}
V_r^{13}	1.44×10^{-3}	4.77×10^{-3}	1.58×10^{-2}	6.06×10^{-2}

6.3.3 Estimation in the presence of colored noise

The robust DSE method has also been tested in the presence of non-Gaussian noises. This testing has been done by including three different

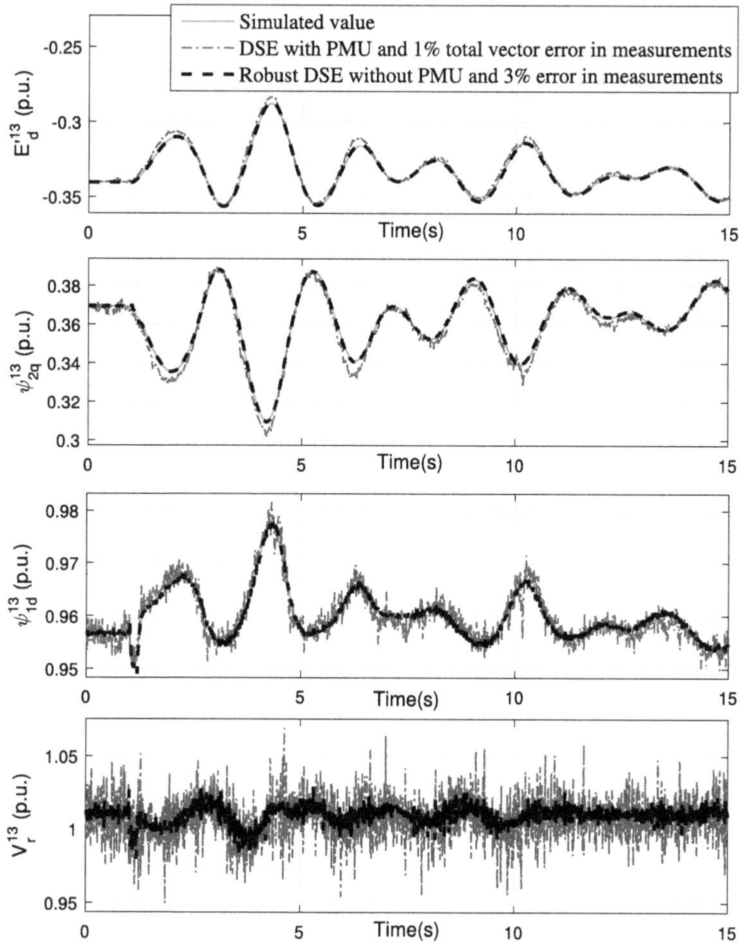

Figure 6.3 Comparison of DSE for $E_d'^{13}$, ψ_{2q}^{13}, ψ_{1d}^{13}, and V_r^{13} for the base case.

colored noises in the measurements: pink noise, blue noise, and violet noise. The estimation results in the presence of colored noises for both the robust DSE method and the DSE-with-PMU method are presented in Fig. 6.6 and Tables 6.3 and 6.4. It can be observed from these figure and tables that the robust DSE method remains robust to non-Gaussian noises as well, while the DSE-with-PMU method gives inaccurate estimation results. It can also be observed that the robust DSE method has higher estimation errors for pink noise than for blue or violet noises.

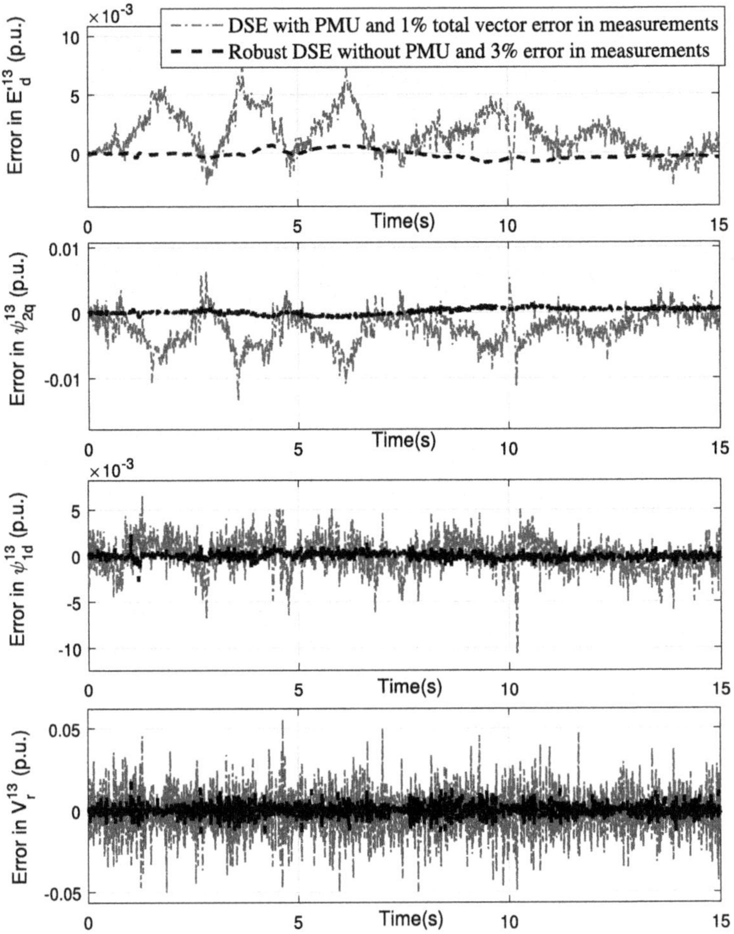

Figure 6.4 Estimation errors for $E_d'^{13}$, ψ_{2q}^{13}, ψ_{1d}^{13}, and V_r^{13} for the base case.

6.3.4 Computational feasibility

Computational feasibility of the robust DSE method can be inferred from the fact that the entire simulation, including simulation of the power system, with two estimators at each machine, runs in faster-than-real-time on MATLAB-Simulink running on Windows 7 on a personal computer with Intel Core 2 Duo, 2.0 GHz CPU and 2 GB RAM. The expression "faster-than-real-time" here means that one second of the simulation takes less than 1 second of processing time. Also, the total execution time for all the operations for the robust DSE method for one time step (that is, for

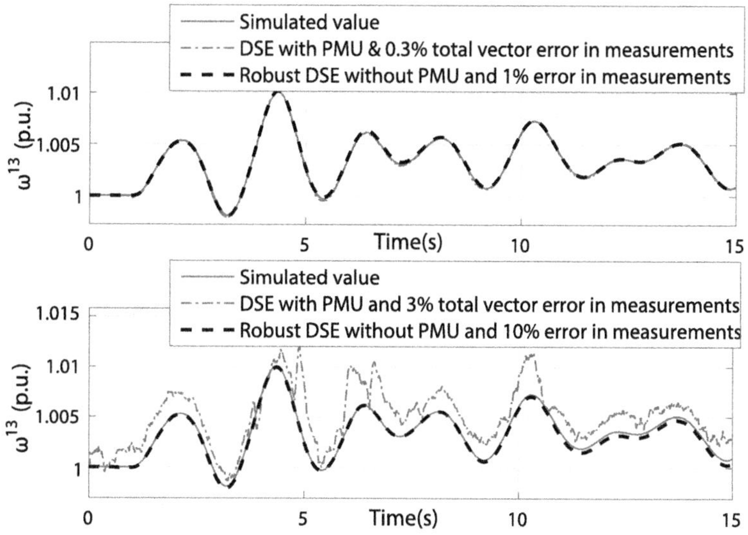

Figure 6.5 Comparison of DSE for ω^{13} for varying noise levels.

Table 6.3 Root mean-square errors for robust DSE with non-Gaussian colored noises

Estimation of RMSEs (in p.u.) for 3% noise in measurements

State	Pink noise	Blue noise	Violet noise
α^{13}	8.86×10^{-4}	4.84×10^{-4}	4.87×10^{-4}
ω^{13}	9.26×10^{-5}	9.73×10^{-5}	9.77×10^{-5}
$E_q'^{13}$	3.54×10^{-4}	1.55×10^{-4}	1.56×10^{-4}
$E_d'^{13}$	5.29×10^{-4}	3.25×10^{-4}	3.27×10^{-4}
ψ_{2q}^{13}	6.88×10^{-4}	3.55×10^{-4}	3.56×10^{-4}
ψ_{1d}^{13}	7.59×10^{-4}	1.92×10^{-4}	1.91×10^{-4}
V_r^{13}	7.92×10^{-3}	5.94×10^{-4}	5.19×10^{-4}

Table 6.4 Root mean-square errors for DSE-with-PMU with non-Gaussian colored noises

Estimation of RMSEs (in p.u.) for 1% noise in measurements

State	Pink noise	Blue noise	Violet noise
α^{13}	6.84×10^{-3}	1.18×10^{-2}	1.85×10^{-2}
ω^{13}	3.89×10^{-4}	3.45×10^{-4}	5.03×10^{-4}
$E_q'^{13}$	9.71×10^{-4}	2.61×10^{-3}	5.02×10^{-3}
$E_d'^{13}$	1.64×10^{-3}	4.23×10^{-3}	6.56×10^{-3}
ψ_{2q}^{13}	2.54×10^{-3}	5.35×10^{-3}	7.98×10^{-3}
ψ_{1d}^{13}	1.99×10^{-3}	2.89×10^{-3}	4.60×10^{-3}
V_r^{13}	1.43×10^{-2}	2.07×10^{-2}	2.69×10^{-2}

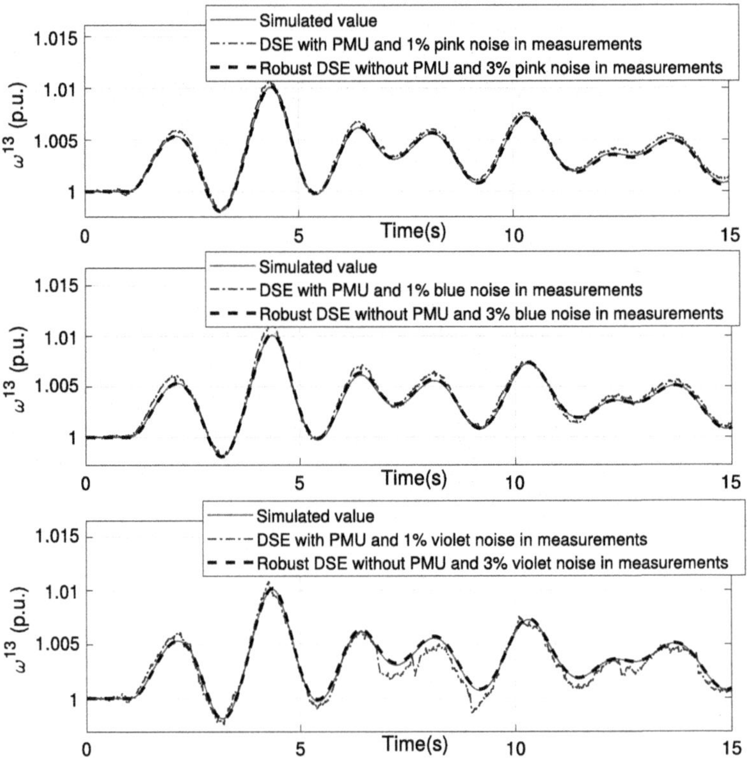

Figure 6.6 Comparison of DSE for ω^{13} for colored noises for the base case.

one iteration) is 0.44 ms. Specifically, execution time for the DFT stage is 0.11 ms, while that for UKF is 0.33 ms (for both the robust DSE method and the DSE-with-PMU method). Thus, the method can be easily implemented using current technologies as the update rate required by the robust DSE method is 10 ms.

6.4 EXTENSION TO AN UNBALANCED SYSTEM

Although the robust DSE method has been developed for a balanced system, it can be extended to an unbalanced three-phase system. Such an extension can be based on applying the interpolated-DFT method to each phase of $V(t)$ and $I(t)$ and finding the magnitude and phase angle of each phase separately. These values can then be transformed using Fortescue transformation to symmetrical components, that is, to positive, negative, and zero-sequence components, and, thereby, the magnitude and phase angle (and hence frequency) of each symmetrical component can be obtained. Finally, the UKF method can be applied to each symmetrical component individually to find the positive, negative, and zero-sequence state estimates for the generation unit.

6.5 SUMMARY

This chapter presents a method for decentralized dynamic state estimation in power systems which works using analogue measurements from instrument transformers to make the estimation robust to time synchronization errors. The presented method is also robust to a wide range of measurement noises which can be encountered in state-of-the-art instrument transformers and has practical computational requirements for real-time operation. The method employs the concept of the internal angle of a generating unit and consists of two stages. In the first stage, signal parameter estimation is performed using interpolated DFT. This is followed by the second stage, in which unscented Kalman filtering is applied to the first-stage estimates to obtain the final state estimates. The advantages of the method over PMU-based decentralized DSE have been presented using the 68-bus test system. To further elaborate, the advantages and features of the presented method are enumerated as follows:

- All the dynamic states are estimated without any time synchronization by incorporating the internal angle in the estimation model, which in turn ensures robustness of the method to synchronization errors.
- The method is considered to remain accurate for varying levels of errors in measurements – from 0.1% to 10%. Also, none of the currently available methods take into account GPS synchronization errors.
- As synchronization is not required for estimation of the states, DSE for these states can be performed using the analogue measurements directly acquired from CTs and VTs. This is particularly beneficial for decentralized control purposes.
- The method consists of a dual-stage estimation process in which interpolated DFT and UKF have been combined as two stages of estimation. The DFT stage dynamically provides estimates of means and variances of the inputs required by the UKF stage, and this continuous updating of variances is one of the reasons for the noise robustness of the method.

CHAPTER 7

Control Based on Dynamic Estimation: Linear and Nonlinear Theories

This chapter aims to provide a basis for decentralized control laws using the dynamic state estimates which were obtained in the previous chapters. A control law should be such that it minimizes the state deviation costs and the required control effort. The theory of optimal control of dynamic systems is appropriate in this context, as it involves cost-effective operation of a system by optimizing the sum of costs associated with system states and control inputs. Linear control is the most developed and widely adopted technology for ensuring small signal stability in power systems (as a nonlinear power system can be approximated with a linear equivalent if deviations from the equilibrium point are small). Nonlinear control is required for ensuring transient stability: when deviations from the equilibrium point are large (due to large disturbances to the system) and linear approximations are no longer valid. As both small signal stability and transient stability are equally important for power system operation, this chapter focuses on both linear and nonlinear optimal control theories.

The rest of the chapter is organized as follows. Section 7.1 focuses on linear optimal control theory, while Section 7.2 briefly describes nonlinear optimal control theory. Section 7.1.1 formally states the linear optimal control problem. Section 7.1.2 explains the classical linear quadratic (LQ) regulator (LQR) solution, while Section 7.1.3 describes the extended linear quadratic (ELQ) problem and its solution. This solution is demonstrated on an example linear time-invariant (LTI) system in Section 7.1.4. Finally, Section 7.3 concludes the chapter.

7.1 LINEAR OPTIMAL CONTROL

The particular case of optimal control in which dynamics of a system are described by linear differential equations and cost is a quadratic function of the states and control effort is called an LQ problem [77,114]. A solution to the LQ problem is provided by LQR, which is a state feedback controller [98].

Dynamic Estimation and Control of Power Systems
https://doi.org/10.1016/B978-0-12-814005-5.00018-2

The decentralized differential and algebraic equations (DAEs) used in the previous chapters involved pseudoinputs. As pseudoinputs are actually measurements, they can neither be disabled nor be manipulated, unlike traditional inputs given to a system. Pseudoinputs are very similar to exogenous inputs found in control literature. Some examples of dynamic systems with exogenous inputs can be found in [115–121]. Pseudoinputs will be referred to as exogenous inputs in this chapter in order to develop a generalized framework for the control of linear systems with such inputs.

The problem of optimal control of LTI systems with exogenous inputs has been termed the ELQ problem in this chapter. It is assumed in the LQR solution that all of the inputs given to the system are normal inputs, which means that each of the input can be manipulated by the controller. The ELQ problem cannot be addressed by the LQR solution, as the exogenous inputs can neither be avoided nor be changed. So, they alter the dynamic behavior of the system and the associated costs. A solution for the ELQ problem will need to incorporate feedback terms corresponding to the exogenous inputs. Let us clearly state the ELQ problem first.

7.1.1 Problem statement

Let us first provide some preliminary definitions.

Definition 7.1. A "control input" given to an LTI system is an input whose magnitude can be decided and changed as per any required control scheme. This is the input in the traditional sense of system theory.

Definition 7.2. An "exogenous input" given to an LTI system is an input which cannot be removed from the system and whose magnitude cannot be decided or changed. This input is an unavoidable quantity which cannot be used as a control input in corrective actions, although it may be possible to find a control input which cancels out or minimizes the effect of the exogenous input.

Using the above two definitions, the control problem is stated as follows:

For a discrete-time open-loop LTI system in which both control inputs and exogenous inputs are present, find an optimal control law such that the sum of the quadratic costs associated with the system state deviations, the exogenous inputs, and the control inputs is minimized.

Thus, the aim of the above problem is to control the system via its control inputs under the constraints of exogenous inputs. This problem is termed the ELQ problem in this chapter.

7.1.2 Classical LQR control

A discrete-time open-loop LTI system without any exogenous input is represented by the following equation:

$$\boldsymbol{x}_{k+1} = \boldsymbol{A}\boldsymbol{x}_k + \boldsymbol{B}\boldsymbol{u}_k. \tag{7.1}$$

The quadratic cost function for (7.1) for $N+1$ samples is given by

$$\boldsymbol{J} = \sum_{k=0}^{N} [\boldsymbol{x}_k^T \mathbf{Q}\boldsymbol{x}_k + \boldsymbol{u}_k^T \boldsymbol{R}\boldsymbol{u}_k], \text{ where } \mathbf{Q} \geq 0, \ \boldsymbol{R} > 0. \tag{7.2}$$

Minimizing $\boldsymbol{J}$ with respect to $\boldsymbol{u}_k$ gives the following LQR solution:

$$\boldsymbol{u}_k = -\boldsymbol{F}_k\boldsymbol{x}_k, \quad k = 0, 1, \ldots, (N-1), \ \boldsymbol{u}_N = \mathbf{0}, \text{ where} \tag{7.3}$$

$$\boldsymbol{F}_{k-1} = (\boldsymbol{R} + \boldsymbol{B}^T\boldsymbol{P}_k\boldsymbol{B})^{-1}\boldsymbol{B}^T\boldsymbol{P}_k\boldsymbol{A}, \ \boldsymbol{P}_N = \mathbf{Q}, \text{ and} \tag{7.4}$$

$$\boldsymbol{P}_{k-1} = \mathbf{Q} + \boldsymbol{A}^T[\boldsymbol{P}_k - \boldsymbol{P}_k\boldsymbol{B}(\boldsymbol{R}+\boldsymbol{B}^T\boldsymbol{P}_k\boldsymbol{B})^{-1}\boldsymbol{B}^T\boldsymbol{P}_k]\boldsymbol{A}. \tag{7.5}$$

If N is finite, then the above optimal control policy is called finite-horizon LQR; otherwise it is infinite-horizon LQR. Moreover, $\boldsymbol{P}_k$ and $\boldsymbol{F}_k$ for the infinite-horizon case are bounded and have a steady-state solution iff the pair $(\boldsymbol{A}, \boldsymbol{B})$ is stabilizable, and the steady-state solution is found by solving the following discrete-time algebraic Riccati equation (ARE). We have

$$\boldsymbol{P} = \mathbf{Q} + \boldsymbol{A}^T[\boldsymbol{P} - \boldsymbol{P}\boldsymbol{B}(\boldsymbol{R} + \boldsymbol{B}^T\boldsymbol{P}\boldsymbol{B})^{-1}\boldsymbol{B}^T\boldsymbol{P}]\boldsymbol{A}, \tag{7.6}$$

$$\boldsymbol{F} = (\boldsymbol{R} + \boldsymbol{B}^T\boldsymbol{P}\boldsymbol{B})^{-1}\boldsymbol{B}^T\boldsymbol{P}\boldsymbol{A}. \tag{7.7}$$

7.1.3 Linear quadratic control for systems with exogenous inputs

A discrete-time open-loop LTI system with both control inputs and exogenous inputs is given by the following equation:

$$\boldsymbol{x}_{k+1} = \boldsymbol{A}\boldsymbol{x}_k + \boldsymbol{B}\boldsymbol{u}_k + \boldsymbol{B}'\boldsymbol{u}_k'. \tag{7.8}$$

As explained in Section 7.1, many solutions have been proposed for linear quadratic control of the above system using disturbance accommodation (such as in [115–121]). In these solutions, the component of the control input which accommodates the exogenous inputs, $\boldsymbol{u}_k'$, is given by $-\boldsymbol{B}^+\boldsymbol{B}'\boldsymbol{u}_k'$.

Here, $\boldsymbol{B}^{+}$ denotes the Moore–Penrose pseudoinverse of $\boldsymbol{B}$. The net control input is given by

$$\boldsymbol{u}_k = -\boldsymbol{F}_k\boldsymbol{x}_k - \boldsymbol{B}^{+}\boldsymbol{B}'\boldsymbol{u}'_k, \text{ where (7.4)–(7.5) give } \boldsymbol{F}_k. \tag{7.9}$$

The above control solution does not consider the quadratic cost for the discrete system given by (7.8). This system has an extra term (corresponding to the exogenous inputs) as compared to the system given by (7.1). Thus the quadratic cost for this system gets modified. For $N+1$ samples it is given by

$$\boldsymbol{J}' = \sum\nolimits_{k=0}^{N} [\boldsymbol{x}_k^T \boldsymbol{Q}\boldsymbol{x}_k + \boldsymbol{u}_k^T \boldsymbol{R}\boldsymbol{u}_k + \boldsymbol{u}_k'^T \boldsymbol{R}'\boldsymbol{u}'_k], \text{ where}$$
$$\boldsymbol{Q} \geq 0, \ \boldsymbol{R} > 0, \ \boldsymbol{R}' \geq 0. \tag{7.10}$$

In order to find the optimal control policy for (7.8), $\boldsymbol{J}'$ in (7.10) needs to be minimized with respect to $\boldsymbol{u}_k$. This minimization gives the following theorem.

Theorem 7.1. *For an LTI system with exogenous inputs or pseudoinputs (as given by* (7.8)*), provided $\boldsymbol{u}'_k = \boldsymbol{0} \ \forall \ k \geq N$, the optimal control policy for $0 \leq k < N$ is given by* (7.11)–(7.14) *(and for $k \geq N$, $\boldsymbol{u}_k = \boldsymbol{0}$). We have*

$$\boldsymbol{u}_k = -(\boldsymbol{F}_k\boldsymbol{x}_k + \boldsymbol{G}_k\boldsymbol{u}'_k + \boldsymbol{G}'_k), \tag{7.11}$$

$$\boldsymbol{G}_k = \boldsymbol{F}_k(\boldsymbol{P}_k - \boldsymbol{Q})^{-1}\boldsymbol{S}_k, \ \boldsymbol{G}'_k = \boldsymbol{F}_k(\boldsymbol{P}_k - \boldsymbol{Q})^{-1}\boldsymbol{S}'_k, \tag{7.12}$$

$$\boldsymbol{S}_N = \boldsymbol{0}, \ \boldsymbol{S}'_N = \boldsymbol{0}, \ \boldsymbol{S}_k = (\boldsymbol{A} - \boldsymbol{B}\boldsymbol{F}_k)^T(\boldsymbol{P}_{k+1}\boldsymbol{B}' + \boldsymbol{S}_{k+1}), \tag{7.13}$$

$$\boldsymbol{S}'_k = (\boldsymbol{A} - \boldsymbol{B}\boldsymbol{F}_k)^T(\boldsymbol{S}_{k+1}(\boldsymbol{u}'_{k+1} - \boldsymbol{u}'_k) + \boldsymbol{S}'_{k+1}), \tag{7.14}$$

where $\boldsymbol{F}_k$ and $\boldsymbol{P}_k$ remain the same as in the LQR case (given by (7.4)–(7.5)*).*

Proof. A preliminary modification needs to be done in the system given by (7.8) for the derivation of Theorem 7.1, by adding a constant exogenous input at the end of the column vector $\boldsymbol{u}'_k$ as

$$\boldsymbol{x}_{k+1} = \boldsymbol{A}\boldsymbol{x}_k + \boldsymbol{B}\boldsymbol{u}_k + \boldsymbol{B}_1\boldsymbol{v}_k, \text{ where, } \boldsymbol{v}_k = \begin{bmatrix} \boldsymbol{u}'_k \\ 1 \end{bmatrix}, \ \boldsymbol{B}_1 = \begin{bmatrix} \boldsymbol{B}' & \boldsymbol{0}_{m\times 1} \end{bmatrix}. \tag{7.15}$$

Also, $\boldsymbol{v}_k = \boldsymbol{E}_k \boldsymbol{v}_{k-1}$,

$$\text{where } \boldsymbol{E}_k = \begin{bmatrix} \boldsymbol{I}_r & \Delta \boldsymbol{u}'_k \\ \boldsymbol{0}_{1\times r} & 1 \end{bmatrix} \text{ and } \Delta \boldsymbol{u}'_k = \boldsymbol{u}'_k - \boldsymbol{u}'_{k-1}, \tag{7.16}$$

$$\boldsymbol{B}_1 \boldsymbol{v}_k = \begin{bmatrix} \boldsymbol{B}' & \boldsymbol{0}_{m\times 1} \end{bmatrix} \begin{bmatrix} \boldsymbol{u}'_k \\ 1 \end{bmatrix} = \boldsymbol{B}' \boldsymbol{u}'_k + \boldsymbol{0}_{m\times 1} = \boldsymbol{B}' \boldsymbol{u}'_k. \tag{7.17}$$

Here m is the number of elements in $\boldsymbol{x}_k$ and r is the number of elements in $\boldsymbol{u}'_k$. It should be understood that because of (7.17), the above modification has no effect on the dynamics of the original system. The modification is needed to get an iterative expression for the optimal control policy. On its own, $\boldsymbol{u}'_k$ cannot be expressed in terms of $\boldsymbol{u}'_{k-1}$. But when a new pseudoinput vector $\boldsymbol{v}_k$ is defined by appending a constant value 1 at the end of $\boldsymbol{u}'_k$, then $\boldsymbol{v}_k$ can be expressed in terms of $\boldsymbol{v}_{k-1}$ using (7.16). The quadratic cost for the modified system (given by (7.15)) for N samples is given by

$$\boldsymbol{J}' = \sum_{k=0}^{N-1} [\boldsymbol{x}_k^T \boldsymbol{Q} \boldsymbol{x}_k + \boldsymbol{u}_k^T \boldsymbol{R} \boldsymbol{u}_k + \boldsymbol{v}_k^T \boldsymbol{R}_1 \boldsymbol{v}_k], \tag{7.18}$$

$$\text{where } \boldsymbol{R}_1 = \begin{bmatrix} \boldsymbol{R}' & \boldsymbol{0}_{r\times 1} \\ \boldsymbol{0}_{1\times r} & 0 \end{bmatrix}, \ \boldsymbol{Q} \geq 0, \ \boldsymbol{R} > 0, \ \boldsymbol{R}' \geq 0, \tag{7.19}$$

$$\boldsymbol{v}_k^T \boldsymbol{R}_1 \boldsymbol{v}_k = \begin{bmatrix} \boldsymbol{u}'^T_k & 1 \end{bmatrix} \begin{bmatrix} \boldsymbol{R}' & \boldsymbol{0}_{r\times 1} \\ \boldsymbol{0}_{1\times r} & 0 \end{bmatrix} \begin{bmatrix} \boldsymbol{u}'_k \\ 1 \end{bmatrix} = \boldsymbol{u}'^T_k \boldsymbol{R}' \boldsymbol{u}'_k. \tag{7.20}$$

Eq. (7.20) and the definition of $\boldsymbol{R}_1$ (given by (7.19)) ensure that the constant exogenous input in $\boldsymbol{v}_k$ has zero cost, so that the quadratic costs for the modified system and the original system (as given by (7.18) and (7.10), respectively) are identical.

As it is given that $\boldsymbol{u}'_k = \boldsymbol{0} \ \forall \ k \geq N$ and as the system reaches its final steady state, $\boldsymbol{x}_N$, at $k = N$, the optimal input required is $\boldsymbol{u}_k = \boldsymbol{0} \ \forall \ k \geq N$. The optimal cost for $k = N$ is therefore $\boldsymbol{J}_N'^{opt} = \boldsymbol{x}_N^T \boldsymbol{Q} \boldsymbol{x}_N = \boldsymbol{x}_N^T \boldsymbol{P}_N \boldsymbol{x}_N$. The combined quadratic cost for $k = N - 1$ and $k = N$, provided that the cost for $k = N$ is optimal (which is $\boldsymbol{J}_N'^{opt}$), is given by $\boldsymbol{J}'_{N-1}$ as

$$\boldsymbol{J}'_{N-1} = \boldsymbol{x}_{N-1}^T \boldsymbol{Q} \boldsymbol{x}_{N-1} + \boldsymbol{u}_{N-1}^T \boldsymbol{R} \boldsymbol{u}_{N-1} + \boldsymbol{v}_{N-1}^T \boldsymbol{R}_1 \boldsymbol{v}_{N-1} + \boldsymbol{J}_N'^{opt}. \tag{7.21}$$

Substituting $\boldsymbol{J}_N^{\prime opt} = \boldsymbol{x}_N^T \boldsymbol{P}_N \boldsymbol{x}_N$ and $\boldsymbol{x}_N = \boldsymbol{A}\boldsymbol{x}_{N-1} + \boldsymbol{B}\boldsymbol{u}_{N-1} + \boldsymbol{B}_1\boldsymbol{v}_{N-1}$ in (7.21), we obtain

$$\boldsymbol{J}'_{N-1} = \boldsymbol{x}_{N-1}^T \mathbf{Q}\boldsymbol{x}_{N-1} + \boldsymbol{u}_{N-1}^T \boldsymbol{R}\boldsymbol{u}_{N-1} + \boldsymbol{v}_{N-1}^T \boldsymbol{R}_1 \boldsymbol{v}_{N-1} + (\boldsymbol{A}\boldsymbol{x}_{N-1} + \boldsymbol{B}\boldsymbol{u}_{N-1} + \boldsymbol{B}_1\boldsymbol{v}_{N-1})^T \boldsymbol{P}_N (\boldsymbol{A}\boldsymbol{x}_{N-1} + \boldsymbol{B}\boldsymbol{u}_{N-1} + \boldsymbol{B}_1\boldsymbol{v}_{N-1}). \tag{7.22}$$

Finding the partial derivative of $\boldsymbol{J}'_{N-1}$ in the above equation with respect to $\boldsymbol{u}_{N-1}$, $\partial \boldsymbol{J}'_{N-1}/\partial \boldsymbol{u}_{N-1}$, we have

$$\partial \boldsymbol{J}'_{N-1}/\partial \boldsymbol{u}_{N-1} = 2[\boldsymbol{R}\boldsymbol{u}_{N-1} + \boldsymbol{B}^T \boldsymbol{P}_N (\boldsymbol{A}\boldsymbol{x}_{N-1} + \boldsymbol{B}\boldsymbol{u}_{N-1} + \boldsymbol{B}_1\boldsymbol{v}_{N-1})], \tag{7.23}$$

$$\because \partial \boldsymbol{J}'_{N-1}/\partial \boldsymbol{u}_{N-1} = \mathbf{0}, \text{ for } \boldsymbol{u}_{N-1} = \boldsymbol{u}_{N-1}^{opt}, \tag{7.24}$$

$$\therefore \boldsymbol{R}\boldsymbol{u}_{N-1}^{opt} + \boldsymbol{B}^T \boldsymbol{P}_N (\boldsymbol{A}\boldsymbol{x}_{N-1} + \boldsymbol{B}\boldsymbol{u}_{N-1}^{opt} + \boldsymbol{B}_1\boldsymbol{v}_{N-1}) = \mathbf{0}, \tag{7.25}$$

$$\Rightarrow \boldsymbol{u}_{N-1}^{opt} = -(\boldsymbol{F}_{N-1}\boldsymbol{x}_{N-1} + \boldsymbol{H}_{N-1}\boldsymbol{v}_{N-1}), \tag{7.26}$$

$$\text{where } \boldsymbol{F}_{N-1} = (\boldsymbol{R} + \boldsymbol{B}^T \boldsymbol{P}_N \boldsymbol{B})^{-1} \boldsymbol{B}^T \boldsymbol{P}_N \boldsymbol{A}, \tag{7.27}$$

$$\boldsymbol{H}_{N-1} = (\boldsymbol{R} + \boldsymbol{B}^T \boldsymbol{P}_N \boldsymbol{B})^{-1} \boldsymbol{B}^T \boldsymbol{P}_N \boldsymbol{B}_1. \tag{7.28}$$

Also, as $\partial^2 \boldsymbol{J}'_{N-1}/(\partial \boldsymbol{u}_{N-1})^2 = (\boldsymbol{R} + \boldsymbol{B}^T \boldsymbol{P}_N \boldsymbol{B}) > 0$ (as $\boldsymbol{R} > 0, \boldsymbol{P}_N \geq 0$) and as $\boldsymbol{J}'_{N-1}$ is a quadratic function of $\boldsymbol{u}_{N-1}$, $\boldsymbol{u}_{N-1}^{opt}$ gives the global minimum for $\boldsymbol{J}'_{N-1}$. Substituting $\boldsymbol{u}_{N-1}^{opt}$ from (7.26) for $\boldsymbol{u}_{N-1}$ in (7.21), we obtain

$$\boldsymbol{J}_{N-1}^{\prime opt} = \boldsymbol{x}_{N-1}^T \boldsymbol{P}_{N-1} \boldsymbol{x}_{N-1} + 2\boldsymbol{x}_{N-1}^T \boldsymbol{U}_{N-1} \boldsymbol{v}_{N-1} + \boldsymbol{v}_{N-1}^T \boldsymbol{W}_{N-1} \boldsymbol{v}_{N-1}, \tag{7.29}$$

$$\text{where } \boldsymbol{P}_{N-1} = \mathbf{Q} + \boldsymbol{F}_{N-1}^T \boldsymbol{R}\boldsymbol{F}_{N-1} + (\boldsymbol{A} - \boldsymbol{B}\boldsymbol{F}_{N-1})^T \boldsymbol{P}_N (\boldsymbol{A} - \boldsymbol{B}\boldsymbol{F}_{N-1}), \tag{7.30}$$

$$\boldsymbol{U}_{N-1} = \boldsymbol{F}_{N-1}^T \boldsymbol{R}\boldsymbol{H}_{N-1} + (\boldsymbol{A} - \boldsymbol{B}\boldsymbol{F}_{N-1})^T \boldsymbol{P}_N (\boldsymbol{B}_1 - \boldsymbol{B}\boldsymbol{H}_{N-1}), \tag{7.31}$$

$$\boldsymbol{W}_{N-1} = \boldsymbol{R}_1 + \boldsymbol{H}_{N-1}^T \boldsymbol{R}\boldsymbol{H}_{N-1} + (\boldsymbol{B}_1 - \boldsymbol{B}\boldsymbol{H}_{N-1})^T \boldsymbol{P}_N (\boldsymbol{B}_1 - \boldsymbol{B}\boldsymbol{H}_{N-1}). \tag{7.32}$$

Again, the combined quadratic cost for $k = (N-2)$, $(N-1)$, and N, provided that the combined cost for $k = (N-1)$ and N is optimal (which is $\boldsymbol{J}_{N-1}^{\prime opt}$), is given by $\boldsymbol{J}'_{N-2} = \boldsymbol{x}_{N-2}^T \mathbf{Q}\boldsymbol{x}_{N-2} + \boldsymbol{u}_{N-2}^T \boldsymbol{R}\boldsymbol{u}_{N-2} + \boldsymbol{v}_{N-2}^T \boldsymbol{R}_1 \boldsymbol{v}_{N-2} + \boldsymbol{J}_{N-1}^{\prime opt}$, and following the same aforementioned steps applied to find $\boldsymbol{J}_{N-1}^{\prime opt}$, we find the following values of $\boldsymbol{u}_{N-2}^{opt}$ and $\boldsymbol{J}'^{opt}_{N-2}$:

$$\boldsymbol{u}_{N-2}^{opt} = -(\boldsymbol{F}_{N-2}\boldsymbol{x}_{N-2} + \boldsymbol{H}_{N-2}\boldsymbol{v}_{N-2}), \tag{7.33}$$

$$\text{where } \boldsymbol{F}_{N-2} = (\boldsymbol{R} + \boldsymbol{B}^T\boldsymbol{P}_{N-1}\boldsymbol{B})^{-1}\boldsymbol{B}^T\boldsymbol{P}_{N-1}\boldsymbol{A}, \tag{7.34}$$

$$\boldsymbol{H}_{N-2} = (\boldsymbol{R} + \boldsymbol{B}^T\boldsymbol{P}_{N-1}\boldsymbol{B})^{-1}\boldsymbol{B}^T(\boldsymbol{P}_{N-1}\boldsymbol{B}_1 + \boldsymbol{U}_{N-1}\boldsymbol{E}_{N-1}), \tag{7.35}$$

$$\boldsymbol{J}'^{opt}_{N-2} = \boldsymbol{x}_{N-2}^T\boldsymbol{P}_{N-2}\boldsymbol{x}_{N-2} + 2\boldsymbol{x}_{N-2}^T\boldsymbol{U}_{N-2}\boldsymbol{v}_{N-2} + \boldsymbol{v}_{N-2}^T\boldsymbol{W}_{N-2}\boldsymbol{v}_{N-2}, \tag{7.36}$$

$$\text{where } \boldsymbol{P}_{N-2} = (\boldsymbol{A} - \boldsymbol{B}\boldsymbol{F}_{N-2})^T\boldsymbol{P}_{N-1}(\boldsymbol{A} - \boldsymbol{B}\boldsymbol{F}_{N-2}) + \boldsymbol{F}_{N-2}^T\boldsymbol{R}\boldsymbol{F}_{N-2} + \boldsymbol{Q}, \tag{7.37}$$

$$\begin{aligned}\boldsymbol{U}_{N-2} = {} & (\boldsymbol{A} - \boldsymbol{B}\boldsymbol{F}_{N-2})^T\boldsymbol{P}_{N-1}(\boldsymbol{B}_1 - \boldsymbol{B}\boldsymbol{H}_{N-2}) \\ & + (\boldsymbol{A} - \boldsymbol{B}\boldsymbol{F}_{N-2})^T\boldsymbol{U}_{N-1}\boldsymbol{E}_{N-1} + \boldsymbol{F}_{N-2}^T\boldsymbol{R}\boldsymbol{H}_{N-2},\end{aligned} \tag{7.38}$$

$$\begin{aligned}\boldsymbol{W}_{N-2} = {} & (\boldsymbol{B}_1 - \boldsymbol{B}\boldsymbol{H}_{N-2})^T[\boldsymbol{P}_{N-1}(\boldsymbol{B}_1 - \boldsymbol{B}\boldsymbol{H}_{N-2}) + \boldsymbol{U}_{N-1}\boldsymbol{E}_{N-1}] \\ & + \boldsymbol{E}_{N-1}^T\boldsymbol{W}_{N-1}\boldsymbol{E}_{N-1} + \boldsymbol{H}_{N-2}^T\boldsymbol{R}\boldsymbol{H}_{N-2} + \boldsymbol{R}_1.\end{aligned} \tag{7.39}$$

Next, when the terms $\boldsymbol{u}_{N-3}^{opt}$ and $\boldsymbol{J}'^{opt}_{N-3}$ are evaluated, their expressions are similar to (7.33) and (7.36), respectively, with the only change that $N-2$ is replaced by $N-3$ and $N-1$ is replaced by $N-2$. Similar expressions come for the rest of $\boldsymbol{u}_k^{opt}$ and $\boldsymbol{J}'^{opt}_k$ (that is, for $k < N-3$). Thus, using initial conditions $\boldsymbol{U}_N = \boldsymbol{0}_{m\times(r+1)}$ and $\boldsymbol{P}_N = \boldsymbol{Q}$ and applying induction for $k < N$, the optimal cost for $\boldsymbol{J}'$ in (7.18) comes as $\boldsymbol{J}'^{opt}_0$ (and is found by iteratively evaluating the sequence $\boldsymbol{J}'^{opt}_N, \boldsymbol{J}'^{opt}_{N-1}, \ldots, \boldsymbol{J}'^{opt}_1, \boldsymbol{J}'^{opt}_0$) and the corresponding optimal control policy required to arrive at this optimal cost is given by

$$\boldsymbol{u}_k^{opt} = -(\boldsymbol{F}_k\boldsymbol{x}_k + \boldsymbol{H}_k\boldsymbol{v}_k), \; 0 \le k < N, \tag{7.40}$$

$$\text{where } \boldsymbol{F}_k = (\boldsymbol{R} + \boldsymbol{B}^T\boldsymbol{P}_{k+1}\boldsymbol{B})^{-1}\boldsymbol{B}^T\boldsymbol{P}_{k+1}\boldsymbol{A}, \tag{7.41}$$

$$\boldsymbol{H}_k = (\boldsymbol{R} + \boldsymbol{B}^T\boldsymbol{P}_{k+1}\boldsymbol{B})^{-1}\boldsymbol{B}^T(\boldsymbol{P}_{k+1}\boldsymbol{B}_1 + \boldsymbol{U}_{k+1}\boldsymbol{E}_{k+1}), \tag{7.42}$$

$$\boldsymbol{P}_k = \boldsymbol{Q} + \boldsymbol{F}_k^T\boldsymbol{R}\boldsymbol{F}_k + (\boldsymbol{A} - \boldsymbol{B}\boldsymbol{F}_k)^T\boldsymbol{P}_{k+1}(\boldsymbol{A} - \boldsymbol{B}\boldsymbol{F}_k), \tag{7.43}$$

$$\boldsymbol{U}_k = \boldsymbol{F}_k^T\boldsymbol{R}\boldsymbol{H}_k + (\boldsymbol{A} - \boldsymbol{B}\boldsymbol{F}_k)^T[\boldsymbol{P}_{k+1}(\boldsymbol{B}_1 - \boldsymbol{B}\boldsymbol{H}_k) + \boldsymbol{U}_{k+1}\boldsymbol{E}_{k+1}]. \tag{7.44}$$

It may be noted that $\boldsymbol{W}_k$ has no role in deciding $\boldsymbol{u}_k^{opt}$. Also, $\boldsymbol{P}_k$ (using (7.43)) can be rewritten as

$$\boldsymbol{P}_k = \mathbf{Q} + \boldsymbol{F}_k^T(\boldsymbol{R} + \boldsymbol{B}^T\boldsymbol{P}_{k+1}\boldsymbol{B})\boldsymbol{F}_k - \boldsymbol{F}_k^T\boldsymbol{B}^T\boldsymbol{P}_{k+1}\boldsymbol{A} + \boldsymbol{A}^T\boldsymbol{P}_{k+1}(\boldsymbol{A} - \boldsymbol{B}\boldsymbol{F}_k), \tag{7.45}$$

$$\because \boldsymbol{F}_k^T(\boldsymbol{R} + \boldsymbol{B}^T\boldsymbol{P}_{k+1}\boldsymbol{B})\boldsymbol{F}_k = \boldsymbol{F}_k^T\boldsymbol{B}^T\boldsymbol{P}_{k+1}\boldsymbol{A} \text{ (from (7.41))}, \tag{7.46}$$

$$\therefore \boldsymbol{P}_k = \mathbf{Q} + \boldsymbol{A}^T\boldsymbol{P}_{k+1}(\boldsymbol{A} - \boldsymbol{B}\boldsymbol{F}_k). \tag{7.47}$$

Substituting $\boldsymbol{F}_k$ from (7.41) in (7.47) gives

$$\boldsymbol{P}_k = \mathbf{Q} + \boldsymbol{A}^T(\boldsymbol{P}_{k+1}\boldsymbol{B}(\boldsymbol{R} + \boldsymbol{B}^T\boldsymbol{P}_{k+1}\boldsymbol{B})^{-1}\boldsymbol{B}^T\boldsymbol{P}_{k+1})\boldsymbol{A}. \tag{7.48}$$

Similarly, $\boldsymbol{U}_k$ (using (7.44)) can be rewritten as

$$\begin{aligned}\boldsymbol{U}_k = {} & (\boldsymbol{A} - \boldsymbol{B}\boldsymbol{F}_k)^T(\boldsymbol{P}_{k+1}\boldsymbol{B}_1 + \boldsymbol{U}_{k+1}\boldsymbol{E}_{k+1}) \\ & + \boldsymbol{F}_k^T(\boldsymbol{R} + \boldsymbol{B}^T\boldsymbol{P}_{k+1}\boldsymbol{B})\boldsymbol{H}_k - \boldsymbol{A}^T\boldsymbol{P}_{k+1}\boldsymbol{B}\boldsymbol{H}_k,\end{aligned} \tag{7.49}$$

$$\because \boldsymbol{F}_k^T(\boldsymbol{R} + \boldsymbol{B}^T\boldsymbol{P}_{k+1}\boldsymbol{B})\boldsymbol{H}_k = \boldsymbol{A}^T\boldsymbol{P}_{k+1}\boldsymbol{B}\boldsymbol{H}_k \text{ (using (7.41))}, \tag{7.50}$$

$$\therefore \boldsymbol{U}_k = (\boldsymbol{A} - \boldsymbol{B}\boldsymbol{F}_k)^T(\boldsymbol{P}_{k+1}\boldsymbol{B}_1 + \boldsymbol{U}_{k+1}\boldsymbol{E}_{k+1}). \tag{7.51}$$

Also, from (7.47),

$$(\boldsymbol{A} - \boldsymbol{B}\boldsymbol{F}_k)^T = (\boldsymbol{P}_k - \mathbf{Q})\boldsymbol{A}^{-1}\boldsymbol{P}_{k+1}^{-1}. \tag{7.52}$$

Substituting $(\boldsymbol{A} - \boldsymbol{B}\boldsymbol{F}_k)^T$ from (7.52) in (7.51), we have

$$\boldsymbol{U}_k = (\boldsymbol{P}_k - \mathbf{Q})\boldsymbol{A}^{-1}(\boldsymbol{B}_1 + \boldsymbol{P}_{k+1}^{-1}\boldsymbol{U}_{k+1}\boldsymbol{E}_{k+1}). \tag{7.53}$$

Using (7.41), $\boldsymbol{H}_k$ in (7.42) can be rewritten as

$$\boldsymbol{H}_k = \boldsymbol{F}_k\boldsymbol{A}^{-1}(\boldsymbol{B}_1 + \boldsymbol{P}_{k+1}^{-1}\boldsymbol{U}_{k+1}\boldsymbol{E}_{k+1}), \tag{7.54}$$

$$\text{and using (7.53)} \Rightarrow \boldsymbol{H}_k = \boldsymbol{F}_k(\boldsymbol{P}_k - \mathbf{Q})^{-1}\boldsymbol{U}_k. \tag{7.55}$$

Partitioning $\boldsymbol{U}_k$ in (7.51) as $\begin{bmatrix}\boldsymbol{S}_k & \boldsymbol{S}'_k\end{bmatrix}$, $\boldsymbol{S}_k \in \mathbb{R}^{m\times r}$, $\boldsymbol{S}'_k \in \mathbb{R}^{m\times 1}$, we have

$$\begin{bmatrix}\boldsymbol{S}_k & \boldsymbol{S}'_k\end{bmatrix} = (\boldsymbol{A} - \boldsymbol{B}\boldsymbol{F}_k)^T(\boldsymbol{P}_{k+1}\boldsymbol{B}_1 + \begin{bmatrix}\boldsymbol{S}_{k+1} & \boldsymbol{S}'_{k+1}\end{bmatrix}\boldsymbol{E}_{k+1}) \tag{7.56}$$

$$\Rightarrow \begin{bmatrix} S_k & S'_k \end{bmatrix} = (A - BF_k)^T (P_{k+1} \begin{bmatrix} B' & 0_{m\times 1} \end{bmatrix}$$

$$+ \begin{bmatrix} S_{k+1} & S'_{k+1} \end{bmatrix} \begin{bmatrix} I_r & \Delta u'_{k+1} \\ 0_{1\times r} & 1 \end{bmatrix}) \tag{7.57}$$

$$\Rightarrow S_k = (A - BF_k)^T (P_{k+1} B' + S_{k+1}) \text{ and} \tag{7.58}$$

$$S'_k = (A - BF_k)^T (S_{k+1}(u'_{k+1} - u'_k) + S'_{k+1}). \tag{7.59}$$

Partitioning H_k in (7.40) as $\begin{bmatrix} G_k & G'_k \end{bmatrix}$, $G_k \in \mathbb{R}^{p\times r}$, $G'_k \in \mathbb{R}^{p\times 1}$, where p is the number of elements in u_k, we have

$$u_k^{opt} = -\left(F_k x_k + \begin{bmatrix} G_k & G'_k \end{bmatrix} \begin{bmatrix} u'_k \\ 1 \end{bmatrix} \right) \tag{7.60}$$

$$\Rightarrow u_k^{opt} = -(F_k x_k + G_k u'_k + G'_k) \tag{7.61}$$

and using (7.55), $\begin{bmatrix} G_k & G'_k \end{bmatrix} = F_k (P_k - Q)^{-1} \begin{bmatrix} S_k & S'_k, \end{bmatrix}$

$$\Rightarrow G_k = F_k (P_k - Q)^{-1} S_k, \; G'_k = F_k (P_k - Q)^{-1} S'_k. \tag{7.62}$$

Hence, with (7.41), (7.48), (7.58)–(7.62), Theorem 7.1 stands proved. □

The optimal control solution in Theorem 7.1 is termed the extended linear quadratic regulator (ELQR) solution. If the pair (A, B) is stabilizable, then infinite-horizon solutions for P_k, F_k, G_k, and S_k exist and are given by F, P as in (7.6)–(7.7), and S, G as in (7.65)–(7.67). We have

$$S = (A - BF)^T (PB' + S) = (P - Q) A^{-1} (B' + P^{-1} S) \tag{7.63}$$

$$\text{(this is because } (A - BF)^T = (P - Q) A^{-1} P^{-1} \text{ from (7.52))} \tag{7.64}$$

$$\Rightarrow S = \left(A (P - Q)^{-1} - P^{-1} \right)^{-1} B', \tag{7.65}$$

$$G = F (P - Q)^{-1} S, \text{ and substituting } S \text{ from (7.65)} \tag{7.66}$$

$$\Rightarrow G = F \left(A - P^{-1} (P - Q) \right)^{-1} B'. \tag{7.67}$$

Although the terms $\boldsymbol{F}_k$ and $\boldsymbol{P}_k$ for the ELQR case remain the same as the LQR case, this needs to be mathematically derived and hence the above derivation is important. The other terms $\boldsymbol{G}_k$ and $\boldsymbol{S}_k$ are independent of the sequence of $\boldsymbol{u}'_k$, and hence they can be easily calculated if $\boldsymbol{A}$, $\boldsymbol{B}$, $\boldsymbol{B}'$, $\boldsymbol{Q}$, and $\boldsymbol{R}$ are known. On the other hand, the terms $\boldsymbol{G}'_k$ and $\boldsymbol{S}'_k$ require the knowledge of the sequence of $\boldsymbol{u}'_k$ for all the future and present samples. Thus, if the sequence of exogenous inputs is not known, only the terms $\boldsymbol{F}_k$, $\boldsymbol{P}_k$, $\boldsymbol{G}_k$, and $\boldsymbol{S}_k$ can be accurately calculated, and the terms $\boldsymbol{G}'_k$ and $\boldsymbol{S}'_k$ can only be estimated/predicted based on the estimated/predicted values of $\boldsymbol{u}'_k$. If $\boldsymbol{u}'_k$ cannot be estimated/predicted, then the term $\boldsymbol{G}'_k$ should be ignored while finding the ELQR policy.

7.1.4 Implementation example: a third-order LTI system

The ELQR control can be implemented on any system whose equations can be reduced to the form given by (7.8). An illustrative example has been presented as follows, in which a simple third-order LTI system is controlled using the ELQR methodology.

The various state space matrices of the test system, the equation of which is given by (7.8), are as follows:

$$\boldsymbol{A}=\begin{bmatrix} 1.4 & 0.2 & -0.1 \\ -0.2 & 0.8 & -0.3 \\ 0.1 & 0.1 & 0.9 \end{bmatrix},\ \boldsymbol{B}=\begin{bmatrix} 0.1 & 0.8 \\ 1.1 & 0.3 \\ 0.9 & 0.5 \end{bmatrix},\ \text{and}\ \boldsymbol{B}'=\begin{bmatrix} 1.2 \\ 0.1 \\ 0.2 \end{bmatrix}.$$

Hence, the above system has three states, two control inputs, and one exogenous input. Initially, all the states and inputs are zero, that is, $\boldsymbol{x}_0=\boldsymbol{0}_{3\times1}$, $\boldsymbol{u}_0=\boldsymbol{0}_{2\times1}$, and $\boldsymbol{u}'_0=0$. The following three cases are considered for the exogenous input.

7.1.4.1 *Known and deterministic model of the exogenous input*

In this case, the exogenous input can be predicted and its model is known. For $k\geq 1$, it is given as follows:

$$\boldsymbol{u}'_k=(0.95)^k,\ \ k\geq 1. \tag{7.68}$$

Although in the above example system a vanishing exogenous input has been given (which vanishes when $k\to\infty$), any other sequence of exogenous input(s) can be given to the system (which may or may not vanish) and subsequently the ELQR solution can be applied.

ELQR policy

As $\boldsymbol{u}'_k$ in (7.68) is an exponentially decreasing function of the time sample and it becomes zero only when $k \to \infty$, the infinite-horizon case of ELQR needs to be used to optimally control this system. Using Eqs. (7.6), (7.7), (7.65), and (7.67) and the cost weighting matrices $\boldsymbol{Q}$ and $\boldsymbol{R}$ as $\boldsymbol{I}_3$ and $\boldsymbol{I}_2$, respectively, the following infinite-horizon values of $\boldsymbol{P}$, $\boldsymbol{F}$, $\boldsymbol{S}$, and $\boldsymbol{G}$ are evaluated (rounded off to two decimal places):

$$\boldsymbol{P} = \begin{bmatrix} 3.60 & 0.55 & -1.26 \\ 0.55 & 2.28 & -2.37 \\ -1.26 & -2.37 & 6.32 \end{bmatrix}, \ \boldsymbol{F} = \begin{bmatrix} -0.42 & 0.11 & 0.60 \\ 1.14 & 0.12 & 0.05 \end{bmatrix},$$

$$\boldsymbol{S} = \begin{bmatrix} 6.11 \\ 5.46 \\ -11.95 \end{bmatrix}, \text{ and } \boldsymbol{G} = \begin{bmatrix} -1.48 \\ 1.62 \end{bmatrix}. \tag{7.69}$$

Also, $\boldsymbol{S}'_k$ can be evaluated by substituting $\boldsymbol{P}$ and $\boldsymbol{S}$ for $\boldsymbol{P}_k$ (or $\boldsymbol{P}_{k+1}$) and $\boldsymbol{S}_k$, respectively, in (7.14) and solving for $\boldsymbol{S}'_k$ iteratively. $\boldsymbol{S}'_k$ is then substituted in (7.12) to find $\boldsymbol{G}'_k$. The final solutions for $\boldsymbol{S}'_k$ and $\boldsymbol{G}'_k$ are given as (rounded off to two decimal places)

$$\boldsymbol{S}'_k = \begin{bmatrix} -0.67 \\ -1.11 \\ 2.46 \end{bmatrix} \times (0.95)^k = \begin{bmatrix} -0.67 \\ -1.11 \\ 2.46 \end{bmatrix} \boldsymbol{u}'_k \text{ and}$$

$$\boldsymbol{G}'_k = \begin{bmatrix} 0.27 \\ -0.0 \end{bmatrix} \boldsymbol{u}'_k. \tag{7.70}$$

Substituting the values of $\boldsymbol{F}$, $\boldsymbol{G}$, and $\boldsymbol{G}'_k$ from (7.69) and (7.70) in (7.11), the optimal control policy for ELQR becomes

$$\begin{aligned} \boldsymbol{u}_k &= -\begin{bmatrix} -0.42 & 0.11 & 0.60 \\ 1.14 & 0.12 & 0.05 \end{bmatrix} \boldsymbol{x}_k - \begin{bmatrix} -1.48 \\ 1.62 \end{bmatrix} \boldsymbol{u}'_k - \begin{bmatrix} 0.27 \\ -0.02 \end{bmatrix} \boldsymbol{u}'_k \\ &= -\begin{bmatrix} -0.42 & 0.11 & 0.60 \\ 1.14 & 0.12 & 0.05 \end{bmatrix} \boldsymbol{x}_k - \begin{bmatrix} -1.21 \\ 1.60 \end{bmatrix} \boldsymbol{u}'_k. \end{aligned} \tag{7.71}$$

Classical LQR policy

The classical LQR control is also applied on the test system for performance comparison with ELQR control, and as only the state feedback gain $\boldsymbol{F}$ is

Table 7.1 Comparison of net quadratic costs for Case A

Quadratic costs (p.u.)	ELQR	DALQR	Classical LQR
State cost ($\sum x_k^T Q x_k$)	11.76	19.17	54.97
Control effort ($\sum u_k^T R u_k$)	33.28	47.05	79.37
Total cost	45.04	66.22	134.34

required for classical LQR and it is same as the state feedback gain for ELQR control, the classical LQR control policy becomes

$$\boldsymbol{u}_k^{LQR} = -\begin{bmatrix} -0.42 & 0.11 & 0.60 \\ 1.14 & 0.12 & 0.05 \end{bmatrix} \boldsymbol{x}_k. \tag{7.72}$$

Disturbance accommodating LQR policy

The disturbance accommodating LQR (DALQR) policy given by (7.9) is also applied on the test system. After substituting the values for $\boldsymbol{F}$, $\boldsymbol{B}$, and $\boldsymbol{B}'$ in (7.9), the DALQR policy becomes

$$\boldsymbol{u}_k^{DALQR} = -\begin{bmatrix} -0.42 & 0.11 & 0.60 \\ 1.14 & 0.12 & 0.05 \end{bmatrix} \boldsymbol{x}_k - \begin{bmatrix} -0.43 \\ 1.49 \end{bmatrix} \boldsymbol{u}'_k. \tag{7.73}$$

Comparison of control performance

The weighted norms of the states and the control inputs (given by $\boldsymbol{x}_k^T \boldsymbol{Q} \boldsymbol{x}_k$ and $\boldsymbol{u}_k^T \boldsymbol{R} \boldsymbol{u}_k$, respectively) can be used as measures of control performance of a control method. These weighted norms are also the quadratic costs associated with the control method for the kth sample, as can be inferred from the constituent terms of $\boldsymbol{J}'$ in (7.10). The cost associated with the exogenous inputs, given by $\boldsymbol{u}_k'^T \boldsymbol{R}' \boldsymbol{u}'_k$, remains independent of the control method. This is because $\boldsymbol{u}'_k$ is not dependent on the control method.

The test system has been simulated in MATLAB, and the weighted norms of states and control inputs have been plotted in Fig. 7.1. It should be noted that $\boldsymbol{x}_k^T \boldsymbol{Q} \boldsymbol{x}_k = \boldsymbol{x}_k^T \boldsymbol{x}_k$ and $\boldsymbol{u}_k^T \boldsymbol{R} \boldsymbol{u}_k = \boldsymbol{u}_k^T \boldsymbol{u}_k$ for the test system.

Table 7.1 presents a comparison of quadratic costs associated with the states and the control inputs for the three methods.

It may be inferred from Fig. 7.1 and Table 7.1 that ELQR is much more efficient than both DALQR and classical LQR in the presence of exogenous inputs, and for the test system the total quadratic cost for the states and the control inputs is reduced by 66.5% as compared to the classical LQR and by 32.0% as compared to DALQR.

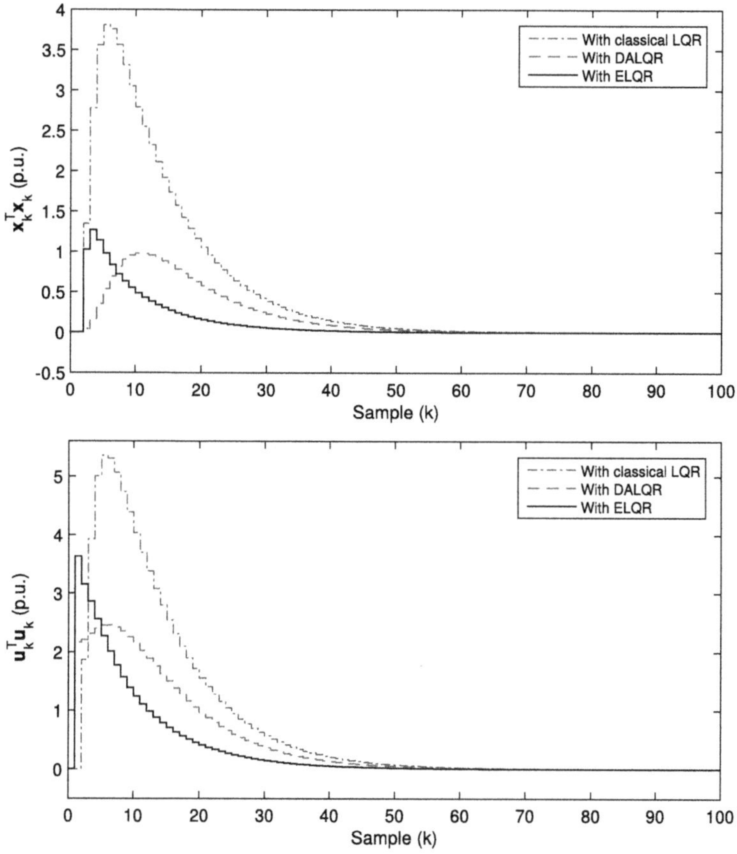

Figure 7.1 Control performance comparison of ELQR with classical LQR and DALQR for Case A.

7.1.4.2 Known and stochastic model of the exogenous input

In this case, the exogenous input is stochastic with the known model, and the rest of the system is the same as in Case A. For $k \geq 1$, the exogenous input is given as follows:

$$\boldsymbol{u}'_k = (X_k)^k, \; X_k \sim N(0.9, 0.01), \; X_k \in (0.85, 0.95). \tag{7.74}$$

Thus, X_k in the above model is a random variable with a truncated normal distribution with mean $= 0.9$, variance $= 0.01$, upper limit $= 0.95$, and lower limit $= 0.85$.

As $\boldsymbol{A}$, $\boldsymbol{B}$, and $\boldsymbol{B}'$ remain the same as in Case A, the values of $\boldsymbol{P}$, $\boldsymbol{F}$, $\boldsymbol{S}$, and $\boldsymbol{G}$ also remain unchanged and are given by (7.69). An exact value of

Table 7.2 Comparison of mean quadratic costs for 1000 simulations for Case B

Mean quadratic costs (p.u.)	ELQR	DALQR	Classical LQR
State cost ($\sum x_k^T Q x_k$)	6.8	9.0	28.5
Control effort ($\sum u_k^T R u_k$)	21.8	24.3	40.8
Total cost	28.6	33.3	69.3

S'_k for this case cannot be found as the sequence of u'_k is nondeterministic. But the expected value of S'_k can be evaluated by substituting $\Delta u'_{k+1}$ with its expected value (which is $-0.1(0.9)^k$) and replacing P_k and S_k with P and S, respectively, in (7.14). Finally, S'_k is solved iteratively and substituted in (7.12) to find G'_k. The final solutions for the expected values of S'_k and G'_k are given as

$$S'_k = \begin{bmatrix} -1.16 \\ -1.87 \\ 4.15 \end{bmatrix} \times (0.9)^k, \quad G'_k = \begin{bmatrix} 0.46 \\ -0.06 \end{bmatrix} \times (0.9)^k. \tag{7.75}$$

Substituting the values of F, G, and G'_k from (7.69) and (7.75) in (7.11), the optimal control policy for ELQR is obtained as follows:

$$u_k = -\begin{bmatrix} -0.42 & 0.11 & 0.60 \\ 1.14 & 0.12 & 0.05 \end{bmatrix} x_k - \begin{bmatrix} -1.48 \\ 1.62 \end{bmatrix} u'_k - \begin{bmatrix} 0.46 \\ -0.06 \end{bmatrix} \times (0.9)^k. \tag{7.76}$$

The optimal control policies for DALQR and classical LQR remain the same as in Case A. As random inputs are used in the simulation, multiple simulations need to be run to get statistics of the quadratic costs. Fig. 7.2 and Table 7.2 show the mean values of quadratic costs for 1000 simulations. It can be observed that, in this case as well, the control performance of ELQR is better as compared to the other two methods, and the net mean quadratic cost for the states and the control inputs is reduced by 58.7% as compared to the classical LQR and by 14.1% as compared to DALQR.

7.1.4.3 Unknown model for the exogenous input

In the third case, it is assumed that any knowledge about the model of the exogenous input is not available, and the exogenous inputs cannot be

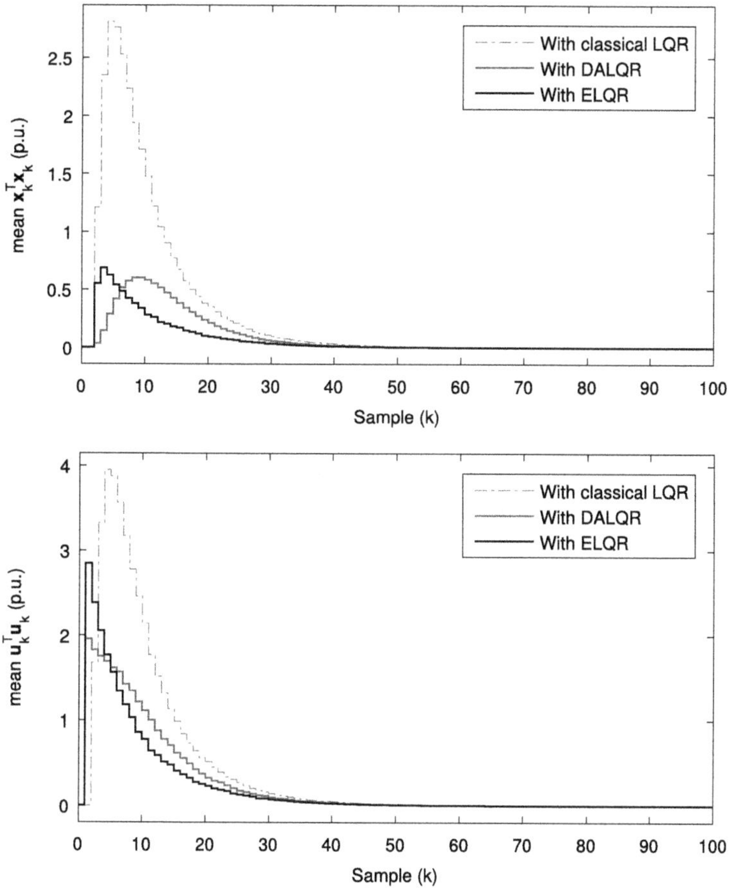

Figure 7.2 Mean control performance for 1000 simulations for Case B.

predicted/estimated. The term $\boldsymbol{G}_k$ in the ELQR policy is not a function of exogenous inputs, and hence it can still be used for control, while the term $\boldsymbol{G}'_k$ must be ignored as it depends on the sequence of exogenous inputs. Substituting the values of $\boldsymbol{F}$ and $\boldsymbol{G}$ from (7.69) in (7.11), the optimal control policy for ELQR becomes as follows, while those for DALQR and classical LQR remain the same as in Case A:

$$\boldsymbol{u}_k = -\begin{bmatrix} -0.42 & 0.11 & 0.60 \\ 1.14 & 0.12 & 0.05 \end{bmatrix} \boldsymbol{x}_k - \begin{bmatrix} -1.48 \\ 1.62 \end{bmatrix} \boldsymbol{u}'_k. \tag{7.77}$$

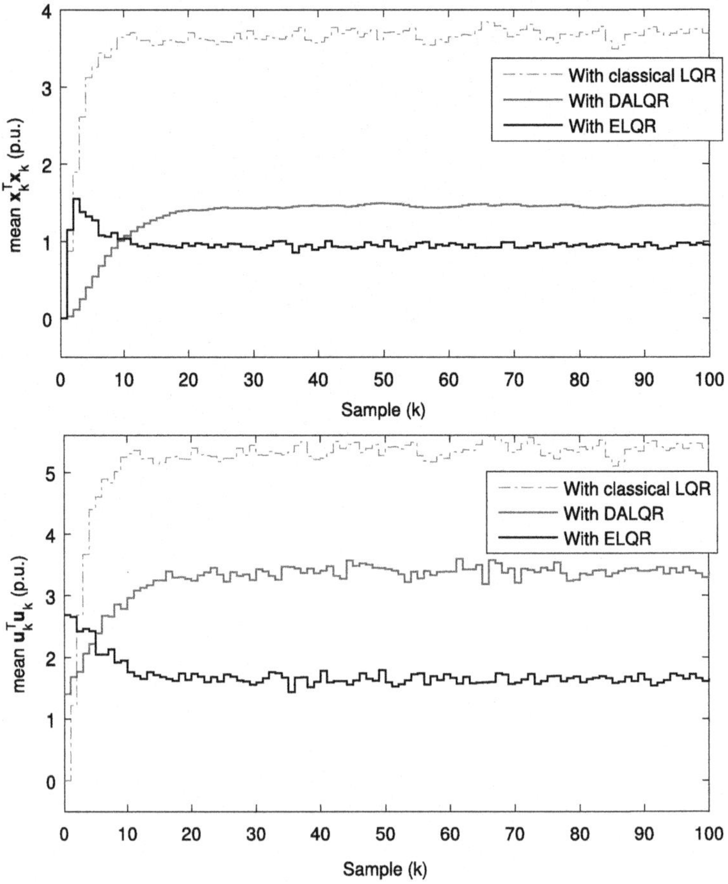

Figure 7.3 Mean control performance for 1000 simulations for Case C.

For simulation, a uniform random number between 0.5 and 1 is given as an exogenous input at each sample (in the above ELQR law, it is assumed that even this information about the randomness of the exogenous inputs is not available). As random inputs are used in this case as well, multiple simulations need to be run to get statistics of the quadratic costs. Fig. 7.3 and Table 7.3 show the mean values of quadratic costs for 1000 simulations. It can be observed that in this case also the control performance of ELQR is much better as compared to the other two methods, and the net mean quadratic cost for the states and the control inputs is reduced by 71.5% as compared to the classical LQR and by 46.9% as compared to DALQR.

Table 7.3 Comparison of mean quadratic costs for 1000 simulations for Case C

Mean quadratic costs (p.u.)	ELQR	DALQR	Classical LQR
State cost ($\sum x_k^T Q x_k$)	93.8	145.4	366.5
Control effort ($\sum u_k^T R u_k$)	163.4	338.6	534.8
Total cost	257.2	484.0	901.3

7.2 NONLINEAR OPTIMAL CONTROL

The nonlinear control methods which have been proposed in the power system literature can broadly be divided into two categories: methods based on normal forms [47–52] and methods based on Lyapunov functions [53–57]. In methods based on normal forms, the dynamics of the system are transformed into a new form (called a normal form) using a curvilinear coordinate system. The transformed system is such that some or all of the transformed states are defined by linear differential equations, and at the same time nonlinearity of the system is exactly preserved in the new form. This transformation process is known as feedback linearization.

In Lyapunov function-based methods, a scalar energy-like function of the system states is found (called Lyapunov function) such that its value is positive at every operating point (except at equilibrium, where it is zero), and the function is nonincreasing along any trajectory of the system.

An advantage of normal form-based methods over Lyapunov-based methods is that a general technique does not exist for finding a Lyapunov function for a system, while the steps for finding a normal form are well established. Moreover, as a normal form is either fully or partially linear, linear control techniques can be used for the linear part of the normal form, whereas linear theory is not applicable in general for Lyapunov-based methods. Owing to these advantages, methods based on normal forms are considered in detail in this book.

7.2.1 Basics of control using normal forms

A general multiinput–multioutput (MIMO) nonlinear system with an equal number of inputs and outputs can be represented in the following form:

$$\dot{x} = f(x) + \sum_{i=1}^{M} g_i(x) u_i, \quad y_i = h_i(x), \quad i = 1, 2, \ldots, M. \tag{7.78}$$

Some preliminary definitions which are required to transform (7.78) to a normal form are provided [122].

Definition 7.3. *Lie derivative*: The Lie derivative of a differentiable scalar function $h(\boldsymbol{x})$ of vector $\boldsymbol{x}$ along a vector field $\boldsymbol{f}(\boldsymbol{x})$, such that $\boldsymbol{f}$ has the same dimension as $\boldsymbol{x}$, is defined as

$$L_f h(\boldsymbol{x}) = \frac{\partial h(\boldsymbol{x})}{\partial \boldsymbol{x}} \boldsymbol{f}(\boldsymbol{x}). \tag{7.79}$$

Definition 7.4. *Relative degree*: For a MIMO nonlinear system given by (7.78), the relative degree of output y_i with respect to the input vector $\boldsymbol{u}$ at a state $\boldsymbol{x}$ is the smallest integer r_i such that $L_{g_j} L_f^{r_i-1} h_i(\boldsymbol{x}) \neq 0$ for at least one $j \in \{1, 2, \ldots, M\}$.

Definition 7.5. *Relative degree set*: A MIMO nonlinear system given by (7.78) has a relative degree set $\{r_1, r_2, \ldots, r_M\}$ at a state $\boldsymbol{x}$ if r_i is the relative degree of y_i, $\forall i \in \{1, 2, \ldots, M\}$, and the $(M \times M)$ characteristic matrix $\boldsymbol{C}(\boldsymbol{x})$, with (i, j) element as $C^{ij}(\boldsymbol{x}) = L_{g_j} L_f^{(r_i-1)} h_i(\boldsymbol{x})$, is nonsingular. If such a set exists, the system is said to have a well-defined relative degree.

With the above three definitions as a base, the following results have been obtained in order to derive a normal form for a MIMO system described by (7.78) (as explained in detail in [122]).

Result 7.1. If a relative degree set $\{r_1, r_2, \ldots, r_M\}$ exists for a system given by (7.78), then $r = r_1 + r_2 + \cdots + r_M \leq n$, where n is the total number of states of the system.

Result 7.2. (a) Provided that $r \leq n$, where $r = r_1 + r_2 + \cdots + r_M$ and r_i is the relative degree of the ith output, define r new states as $z_j^i = \phi_j^i(\boldsymbol{x}) = L_f^{(j-1)} h_i(\boldsymbol{x})$, where, for each $i \in \{1, 2, \ldots, M\}$, j is such that $1 \leq j \leq r_i$. Define $\boldsymbol{z}^i = [z_1^i \; z_2^i \ldots z_{r_i}^i]^T$ and $\boldsymbol{z} = [\boldsymbol{z}^{1^T} \; \boldsymbol{z}^{2^T} \ldots \boldsymbol{z}^{M^T}]^T$. The dynamics for each state of the new state vector $\boldsymbol{z}$ are given as follows (for $1 \leq i \leq M$) and are known as *linearized dynamics* of the system:

$$\begin{aligned} & z_1^i = h_i(\boldsymbol{x}), \;\; \dot{z}_j^i = L_f^j h_i(\boldsymbol{x}) = z_{j+1}^i, \;\; 1 \leq j \leq r_i - 1, \\ & \quad \dot{z}_{r_i}^i = L_f^{r_i} h_i(\boldsymbol{x}) + \textstyle\sum_{j=1}^M L_{g_j} L_f^{(r_i-1)} h_i(\boldsymbol{x}) u_j = v_i \end{aligned} \tag{7.80}$$

$\Rightarrow \boldsymbol{v} = \boldsymbol{D}(\boldsymbol{x}) + \boldsymbol{C}(\boldsymbol{x})\boldsymbol{u}$; $\boldsymbol{D}$, $\boldsymbol{C}$ are as in the nomenclature.

(b) If $r < n$, then define another $(n - r)$ new states, $w_i = \phi_i(\boldsymbol{x})$, where $(r+1) \leq i \leq n$, such that the nonlinear differentiable mapping of the original states to the new states,

$$\boldsymbol{\phi}(\boldsymbol{x}) = [\phi_1^1(\boldsymbol{x}) \ldots \phi_{r_1}^1(\boldsymbol{x}) \ldots \phi_1^M(\boldsymbol{x}) \ldots \phi_{r_M}^M(\boldsymbol{x})\ \phi_{r+1}(\boldsymbol{x}) \ldots \phi_n(\boldsymbol{x})]^T,$$

has a corresponding differentiable inverse mapping, $\boldsymbol{\phi}^{-1}$. That is, if $\boldsymbol{\phi}(\boldsymbol{x}) = (\boldsymbol{z}, \boldsymbol{w})$, then $\boldsymbol{x} = \boldsymbol{\phi}^{-1}(\boldsymbol{z}, \boldsymbol{w})$. It is always possible to find such a mapping for a nonlinear system of form (7.78) with well-defined relative degree. Define $\boldsymbol{w} = [w_{r+1}\ w_{r+2} \ldots w_n]^T$. The dynamics for each state of the state vector $\boldsymbol{w}$ can be written as follows (for $(r+1) \leq i \leq n$) and are known as *internal dynamics* of the system:

$$\begin{aligned} \dot{w}_i &= [L_f \phi_i(\boldsymbol{x}) + \textstyle\sum_{j=1}^{M} L_{g_j} \phi_i(\boldsymbol{x}) u_j]_{\boldsymbol{x}=\boldsymbol{\phi}^{-1}(\boldsymbol{z},\boldsymbol{w})} \\ \Rightarrow \dot{\boldsymbol{w}} &= \boldsymbol{q}(\boldsymbol{z}, \boldsymbol{w}) + \boldsymbol{p}(\boldsymbol{z}, \boldsymbol{w})\boldsymbol{u}; \quad \boldsymbol{q}, \boldsymbol{p} \text{ are as in the nomenclature.} \end{aligned} \tag{7.81}$$

(c) The linearized dynamics and internal dynamics (given by (7.80) and (7.81), respectively) represent the system's normal form.

A simple interpretation of (7.80) is that the output $h_i(\boldsymbol{x})$ is repeatedly differentiated with respect to time (r_i times, to be exact) until the input $\boldsymbol{u}$ appears. After this, $h_i(\boldsymbol{x})$ and its time derivatives $\frac{d^j h_i(\boldsymbol{x})}{dt^j} = L_f^j h_i(\boldsymbol{x})$, $1 \leq j \leq r_i - 1$, are denoted as a new state vector $\boldsymbol{z}^i$ and their dynamics are in a linear form. Thus, this process is known as feedback linearization and the linearized dynamics can be controlled using a linear gain of $\boldsymbol{z}^i$ as a state feedback for input v_i. But, in such a feedback, the system's internal dynamics are unobservable and may be unstable. Thus, for the overall stability of the system, it is necessary that, besides the linearized dynamics, the internal dynamics also remain stable for a given feedback control. The following result formally states the stability criterion [122].

Result 7.3. A MIMO system given by (7.78), with a normal form given by (7.80)–(7.81), is asymptotically stable for a given initial condition and a given feedback control if the closed-loop linearized dynamics are asymptotically stable (that is, the closed-loop poles are in the left half plane) and the internal dynamics are asymptotically stable. If the closed-loop linearized dynamics are asymptotically stable, the asymptotic stability of internal dynamics is equivalent to the asymptotic stability of *zero dynamics*, which are derived from (7.81) by representing $\boldsymbol{u}$ in terms of $\boldsymbol{v}$ (using (7.80), with

$\boldsymbol{x} = \boldsymbol{\phi}^{-1}(\boldsymbol{z}, \boldsymbol{w}))$ and by setting $\boldsymbol{z}$ and $\boldsymbol{v}$ equal to zero. Thus, zero dynamics are given as follows:

$$\dot{\boldsymbol{w}} = \boldsymbol{q}(0, \boldsymbol{w}) - \boldsymbol{p}(0, \boldsymbol{w})\left(\boldsymbol{C}(\boldsymbol{\phi}^{-1}(0, \boldsymbol{w}))\right)^{-1}\boldsymbol{D}(\boldsymbol{\phi}^{-1}(0, \boldsymbol{w})). \tag{7.82}$$

7.3 SUMMARY

This chapter presents various linear and nonlinear optimal control theories which can be used for power system control. A control scheme has been presented for the optimal control of a special case of LTI systems in which both normal and exogenous inputs are present. The scheme is termed extended LQR, and it is shown to be significantly more cost-effective than other LQR schemes. The applicability of the scheme has been shown on a simple model LTI system.

Nonlinear control theories which are used for power systems have also been briefly discussed, with the focus on normal form-based nonlinear control. Most important definitions and results of normal form-based control have been discussed.

CHAPTER 8

Decentralized Linear Control Using DSE and ELQR

This chapter integrates the decentralized dynamic state estimation (DSE) and the extended linear quadratic regulator (ELQR) control scheme developed in Chapters 4, 6 and 7 and uses them to control and to provide adequate damping to the small-signal oscillatory dynamics observed in power systems. The integrated control scheme is completely decentralized. It is a practical alternative to the centralized approach to dynamic system identification and control.

The centralized approach of control of power systems using traditional linear quadratic regulator (LQR) control schemes has been studied in [58–61]. However, this approach, just like every other centralized approach, needs a strong and fast communication network to transmit information and data to the control center. As power systems still lack in such a communication infrastructure (as elaborated in the first two chapters), the centralized approach remains to be more theoretical than practically useful. This chapter aims to address this limitation using the integrated decentralized control scheme.

The rest of the chapter is organized as follows. Section 8.1 describes the architecture of the problem formulation. Section 8.2 explains the concept used for decentralization. The control methodology is detailed in Section 8.3 and Section 8.4 describes the results on a power system model. Section 8.5 summarizes the chapter.

8.1 ARCHITECTURE OF CONTROL

Electromechanical oscillations in power networks are global in nature as they involve large numbers of generators and loads and a significant part of the network. As every generator contributes to these oscillations to varying degrees, each of them can provide suitable control to dampen them out. In the architecture of control, the dynamic states that are obtained for every individual generator (using local phasor measurement unit [PMU] measurements or current/voltage transformer [CT/VT] measurements) are utilized to design a controller that contributes to the overall damping of the

Dynamic Estimation and Control of Power Systems
https://doi.org/10.1016/B978-0-12-814005-5.00019-4

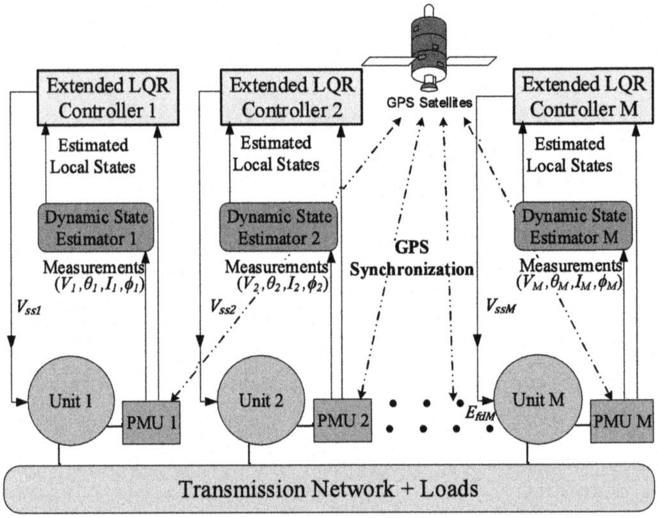

Figure 8.1 Overview of the system and the methodology with PMU.

system-wide oscillations besides any local oscillation. The combined efforts of all the decentralized controllers must produce the desired response of the system at all operating conditions.

An overview of the complete system is given in Fig. 8.1 and Fig. 8.2. In this architecture, each machine can either have both CT/VT and PMU at its terminal that feed voltage and current phasors to the dynamic state estimator which works on the algorithm presented in Chapter 4 – as shown in Fig. 8.1 – or it can have just a pair of CT/VT at its terminal that feed voltage and current analog signals to the dynamic state estimator which works on the algorithm presented in Chapters 5 and 6 – as shown in Fig. 8.2. The state estimates and the measurements are then sent to the local controller, which works on a modified version of the ELQR algorithm (presented in Chapter 7) to calculate an optimal control signal for the excitation system, which in turn controls the excitation of the machine, thereby closing the control loop. The control gains are updated after a small interval (say after every second), so that the control law remains robust to any operating condition. Functionally, even though the dynamic state estimator and the controller are two components, they can be implemented in the same location. The output from the power system stabilizer (PSS) can also be combined with the output of ELQR, but it is not required as such. It should be understood here that a PSS is not necessary when there is an

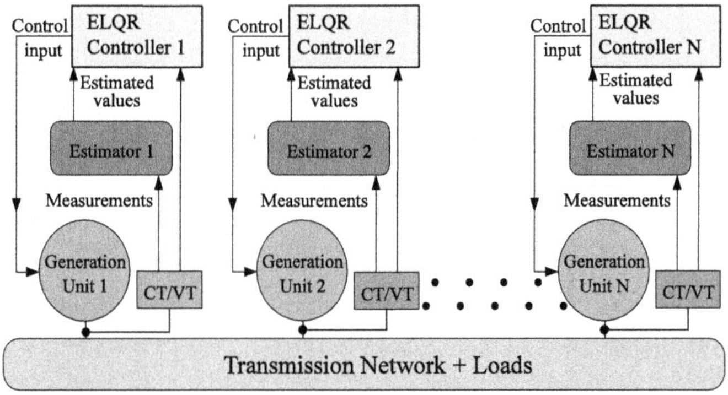

Figure 8.2 Overview of the system and the methodology without PMU.

ELQR in the system, and hence an ELQR can completely substitute a PSS.

It may be noted here that the ELQR controller behaves like a PSS as its output signal directly controls the excitation system of the machine, but there is a fundamental difference between the two. The control gains of the ELQR controller are updated in real-time so that the controller works for any operating point of the system, while the control gain and phase compensator time constants for the PSS are obtained offline for a particular operating condition (or a finite set of operating conditions) using model-based design and then implemented through electronic hardware and/or software.

If PMUs are used for dynamic estimation, then they form an essential part of the DSE-ELQR scheme of estimation and control, and although their functioning has been briefly described in Chapter 4 in Sections 4.1.2 and 4.5.1.1, their effects on control performance still need to be considered. They have been described as follows.

- *Effects of PMU dynamics on controller performance:* The dynamic response of a PMU depends on the combined response of its constituent components, which are the analogue and digital filters and the sampler. The waveforms produced by instrument transformers are processed by the filters for surge suppression and antialiasing filtering in order to filter out high-frequency transients generated during faults and switching operations. There is also an issue of possible aliasing effects due to inadequate sampling rates of the sampler for higher swing frequencies in the network. This issue is rectified by using a decimation filter or a simple averaging filter. Using these functions, PMUs measure the phasors accurately (provided the instrument transformers and GPS satellites are accurate) for both oscillatory and steady-state modes of operation for all practical power systems [16]. Thus, PMU dynamics have no effect on controller performance.

- *Effects of PMU accuracy on controller performance:* The accuracy of PMUs is dependent on the instrument transformers and GPS satellites on which they rely for waveform acquisition and time synchronization, respectively. The waveforms provided by the instrument transformers have errors in both magnitude and phase, but the error in phase can be accurately compensated and calibrated out using digital signal processing (DSP) techniques [100]. Hence the errors in phase are limited only by the time synchronization accuracy of the GPS. The errors in magnitude of these phasors are limited by the accuracy class of the instrument transformers used. These errors in phasors obtained by PMUs can be represented by noises of finite variances, and large errors are considered bad data. These noises and bad data can be filtered out from the dynamic state estimates in the state estimation stage, as shown in Chapter 4, and they have a negligible effect on controller performance, as also demonstrated in Section 8.4.6.

8.2 DECENTRALIZATION OF CONTROL

As explained in Chapter 2, the dynamic behavior of a power system is modeled using a set of continuous-time nonlinear differential and algebraic equations (DAEs). A central control scheme which tunes itself in real-time requires complete knowledge of various states, inputs, and algebraic variables, besides the knowledge of the DAEs. Obtaining such information centrally in real-time is very difficult. However, the local states for the generation unit can be obtained locally in real-time using decentralized DSE. The equation for a single unit is written in a standard form as

$$\dot{\boldsymbol{x}}_{ci}(t) = \boldsymbol{g}_i(\boldsymbol{x}_{ci}(t), \boldsymbol{u}_{ci}(t), \boldsymbol{u}'_{ci}(t)), \text{ where } \boldsymbol{u}_{ci} = V_{ssi}, \tag{8.1}$$

$$\boldsymbol{x}_{ci} = [\delta_i \; \omega_i \; E'_{di} \; E'_{qi} \; \Psi_{1di} \; \Psi_{2qi} \; E'_{dci} \; V_{ri}]^T, \; \boldsymbol{u}'_{ci} = [V_i \; \theta_i]^T.$$

The subscript c in the above equation stands for continuous time. The subscript i in the above equation stands for the ith generation unit. The term $\boldsymbol{u}_{ci}$, which constitutes V_{ssi}, represents the control input to the excitation system of the ith unit and $\boldsymbol{u}'_{ci}$, which constitutes V_i and θ_i (the stator voltage phasors), is the pseudoinput vector for the ith unit, as explained in Chapter 4.

The dynamic model given by (8.1) is formulated considering a reference angle for the system, as a power system is a rotational system. Thus, each δ_i and θ_i is defined with respect to a suitable reference angle, which can either be the rotor angle of a particular reference machine or the center-of-inertia angle, δ_{COI}. This fact is also mentioned in Chapter 6 and illustrated in detail in [62]. But incorporating such a reference angle in the machine model would require the knowledge of the rotor angle of the reference machine (or worse, the knowledge of the rotor angles of all the machines, in the case of δ_{COI}) at each decentralized location, and would therefore defeat the purpose of decentralization. A way of dealing with this problem is discussed in Chapter 6, where a new state $\alpha_i = (\delta_i - \theta_i)$ is defined. As δ_i and θ_i have a common reference angle, it is canceled out in the definition of α_i. The dynamic equation of α_i is given by

$$\dot{\alpha}_i = (\dot{\delta}_i - \dot{\theta}_i) = \omega_B(\omega_i - f_i). \tag{8.2}$$

After incorporating α_i in $\boldsymbol{x}_{ci}$ in (8.1) and replacing the pseudoinput θ_i with its time derivative in p.u., f_i, (8.1) is redefined as

$$\dot{\boldsymbol{x}}_{ci}(t) = \boldsymbol{g}_i(\boldsymbol{x}_{ci}(t), \boldsymbol{u}_{ci}(t), \boldsymbol{u}'_{ci}(t)), \text{ where } \boldsymbol{u}_{ci} = V_{ssi}, \tag{8.3}$$

$$\boldsymbol{x}_{ci} = [\alpha_i\ \omega_i\ E'_{di}\ E'_{qi}\ \Psi_{1di}\ \Psi_{2qi}\ E'_{dci}\ V_{ri}]^T,\ \boldsymbol{u}'_{ci} = [V_i\ f_i]^T. \tag{8.4}$$

The nonlinear equation given by (8.3) needs to be linearized before it can be used in a linear controller. Linearizing (8.3) about an operating point given by $(\boldsymbol{x}_{ci}(t_0), \boldsymbol{u}_{ci}(t_0), \boldsymbol{u}'_{ci}(t_0))$, we have

$$\Delta\dot{\boldsymbol{x}}_{ci}(t) = \boldsymbol{A}_{ci}\Delta\boldsymbol{x}_{ci}(t) + \boldsymbol{B}_{ci}\Delta\boldsymbol{u}_{ci}(t) + \boldsymbol{B}'_{ci}\Delta\boldsymbol{u}'_{ci}(t), \tag{8.5}$$

$$\text{where } \boldsymbol{A}_{ci} = \frac{\partial \boldsymbol{g}_i(t_0)}{\partial \boldsymbol{x}_{ci}(t_0)},\ \boldsymbol{B}_{ci} = \frac{\partial \boldsymbol{g}_i(t_0)}{\partial \boldsymbol{u}_{ci}(t_0)},\ \boldsymbol{B}'_{ci} = \frac{\partial \boldsymbol{g}_i(t_0)}{\partial \boldsymbol{u}'_{ci}(t_0)}, \tag{8.6}$$

$$\Delta\boldsymbol{x}_{ci}(t) = \boldsymbol{x}_{ci}(t) - \boldsymbol{x}_{ci}(t_0),\ \Delta\boldsymbol{u}_{ci}(t) = \boldsymbol{u}_{ci}(t) - \boldsymbol{u}_{ci}(t_0), \tag{8.7}$$

$$\Delta\boldsymbol{u}'_{ci}(t) = \boldsymbol{u}'_{ci}(t) - \boldsymbol{u}'_{ci}(t_0), \text{ and } t \geq t_0. \tag{8.8}$$

Remark. It should be understood that (8.5) remains valid for any operating point of the system, as long as the operating point remains close to an equilibrium point. As (8.5) is used in calculating the ELQR control gains (as explained in subsequent sections), the ELQR control gains also remain valid for each operating point of the system that falls under small-signal dynamic behavior of power systems. The only exception to this fact takes place during a contingency (such as a system fault) during which some of the system states may become transiently unbounded and the system equations can no longer be linearized. Therefore, before linearization and update of control gains it should be checked whether every machine state is within safe operating limits and if not, control gains from the previous sample should be used.

Chapter 2 gives the details of the differential and algebraic functions which constitute $\boldsymbol{g}_i$ for a generating unit. The type of excitation systems used for all the generating units is the same (specifically, static ST1A excitation systems have been used), so that each unit in the system contributes similarly to the net control effort. The details of $\boldsymbol{A}_{ci}$, $\boldsymbol{B}_{ci}$, and $\boldsymbol{B}'_{ci}$ in (8.5) are obtained as follows.

8.2.1 Details of state matrices used in integrated ELQR

Subscripts c and i have been dropped in equations (8.9)–(8.75) for simplicity. This gives

$$\Delta\dot{\boldsymbol{x}} = \boldsymbol{A}\Delta\boldsymbol{x} + \boldsymbol{B}\Delta\boldsymbol{u} + \boldsymbol{B}'\Delta\boldsymbol{u}', \tag{8.9}$$

$$\text{where } \boldsymbol{A} = \frac{\partial \boldsymbol{g}}{\partial \boldsymbol{x}}, \quad \boldsymbol{B} = \frac{\partial \boldsymbol{g}}{\partial \boldsymbol{u}}, \quad \boldsymbol{B}' = \frac{\partial \boldsymbol{g}}{\partial \boldsymbol{u}'}. \tag{8.10}$$

Next, some intermediate partial derivatives are calculated using the DAEs described in Chapter 2. We have

$$\frac{\partial I_d}{\partial \alpha} = \frac{V(R_a \cos\alpha - X''_d \sin\alpha)}{Z_a^2}, \tag{8.11}$$

$$\frac{\partial I_q}{\partial \alpha} = \frac{V(R_a \sin\alpha + X''_d \cos\alpha)}{Z_a^2}, \tag{8.12}$$

$$\frac{\partial I_d}{\partial E'_d} = \frac{R_a K_{q1}}{Z_a^2}, \tag{8.13}$$

$$\frac{\partial I_q}{\partial E'_d} = \frac{X''_d K_{q1}}{Z_a^2}, \tag{8.14}$$

$$\frac{\partial I_d}{\partial E'_q} = \frac{-X''_d K_{d1}}{Z_a^2}, \tag{8.15}$$

$$\frac{\partial I_q}{\partial E'_q} = \frac{R_a K_{d1}}{Z_a^2}, \tag{8.16}$$

$$\frac{\partial I_d}{\partial \Psi_{1d}} = \frac{-X''_d K_{d2}}{Z_a^2}, \tag{8.17}$$

$$\frac{\partial I_q}{\partial \Psi_{1d}} = \frac{R_a K_{d2}}{Z_a^2}, \tag{8.18}$$

$$\frac{\partial I_d}{\partial \Psi_{2q}} = \frac{-R_a K_{q2}}{Z_a^2}, \tag{8.19}$$

$$\frac{\partial I_q}{\partial \Psi_{2q}} = \frac{-X''_d K_{q2}}{Z_a^2}, \tag{8.20}$$

$$\frac{\partial I_d}{\partial E'_{dc}} = \frac{R_a}{Z_a^2}, \tag{8.21}$$

$$\frac{\partial I_q}{\partial E'_{dc}} = \frac{X''_d}{Z_a^2}, \tag{8.22}$$

$$\frac{\partial T_e}{\partial I_d} = E'_d K_{q1} - \Psi_{2q} K_{q2} - I_q(X''_d - X''_q), \tag{8.23}$$

$$\frac{\partial T_e}{\partial I_q} = E'_q K_{d1} - \Psi_{1d} K_{d2} - I_d(X''_d - X''_q), \tag{8.24}$$

$$\frac{\partial I_d}{\partial V} = \frac{R_a \sin\alpha + X''_d \cos\alpha}{Z_a^2}, \tag{8.25}$$

$$\frac{\partial I_q}{\partial V} = \frac{-R_a \cos\alpha + X''_d \sin\alpha}{Z_a^2}. \tag{8.26}$$

Using (8.9), the DAEs described in Chapter 2, and the above intermediate derivatives, the various nonzero terms of $\boldsymbol{A}$, $\boldsymbol{B}$, and $\boldsymbol{B}'$ are given as

$$\boldsymbol{A}_{1,2} = \frac{\partial g_\alpha}{\partial \omega} = \omega_B, \tag{8.27}$$

$$A_{2,1} = \frac{\partial g_\omega}{\partial \alpha} = \frac{-1}{2H}\left(\frac{\partial T_e}{\partial I_d}\frac{\partial I_d}{\partial \alpha} + \frac{\partial T_e}{\partial I_q}\frac{\partial I_q}{\partial \alpha}\right), \tag{8.28}$$

$$A_{2,2} = \frac{\partial g_\omega}{\partial \omega} = \frac{-D}{2H}, \tag{8.29}$$

$$A_{2,3} = \frac{\partial g_\omega}{\partial E'_d} = \frac{-1}{2H}\left(I_d K_{q1} + \frac{\partial T_e}{\partial I_d}\frac{\partial I_d}{\partial E'_d} + \frac{\partial T_e}{\partial I_q}\frac{\partial I_q}{\partial E'_d}\right), \tag{8.30}$$

$$A_{2,4} = \frac{\partial g_\omega}{\partial E'_q} = \frac{-1}{2H}\left(I_q K_{d1} + \frac{\partial T_e}{\partial I_d}\frac{\partial I_d}{\partial E'_q} + \frac{\partial T_e}{\partial I_q}\frac{\partial I_q}{\partial E'_q}\right), \tag{8.31}$$

$$A_{2,5} = \frac{\partial g_\omega}{\partial \Psi_{1d}} = \frac{-1}{2H}\left(I_q K_{d2} + \frac{\partial T_e}{\partial I_d}\frac{\partial I_d}{\partial \Psi_{1d}} + \frac{\partial T_e}{\partial I_q}\frac{\partial I_q}{\partial \Psi_{1d}}\right), \tag{8.32}$$

$$A_{2,6} = \frac{\partial g_\omega}{\partial \Psi_{2q}} = \frac{-1}{2H}\left(-I_d K_{q2} + \frac{\partial T_e}{\partial I_d}\frac{\partial I_d}{\partial \Psi_{2q}} + \frac{\partial T_e}{\partial I_q}\frac{\partial I_q}{\partial \Psi_{2q}}\right), \tag{8.33}$$

$$A_{2,7} = \frac{\partial g_\omega}{\partial E'_{dc}} = \frac{-1}{2H}\left(\frac{\partial T_e}{\partial I_d}\frac{\partial I_d}{\partial E'_{dc}} + \frac{\partial T_e}{\partial I_q}\frac{\partial I_q}{\partial E'_{dc}}\right), \tag{8.34}$$

$$A_{3,1} = \frac{\partial g_{E'_d}}{\partial \alpha} = \frac{-1}{T'_{q0}}\left((X_q - X'_q)K_{q1}\frac{\partial I_q}{\partial \alpha}\right), \tag{8.35}$$

$$A_{3,3} = \frac{\partial g_{E'_d}}{\partial E'_d} = \frac{-1}{T'_{q0}}\left(1 + (X_q - X'_q)\left(K_{q1}\frac{\partial I_q}{\partial E'_d} + \frac{K_{q2}}{X'_q - X_l}\right)\right), \tag{8.36}$$

$$A_{3,4} = \frac{\partial g_{E'_d}}{\partial E'_q} = \frac{-1}{T'_{q0}}\left((X_q - X'_q)K_{q1}\frac{\partial I_q}{\partial E'_q}\right), \tag{8.37}$$

$$A_{3,5} = \frac{\partial g_{E'_d}}{\partial \Psi_{1d}} = \frac{-1}{T'_{q0}}\left((X_q - X'_q)K_{q1}\frac{\partial I_q}{\partial \Psi_{1d}}\right), \tag{8.38}$$

$$A_{3,6} = \frac{\partial g_{E'_d}}{\partial \Psi_{2q}} = \frac{-1}{T'_{q0}}(X_q - X'_q)\left(K_{q1}\frac{\partial I_q}{\partial \Psi_{2q}} + \frac{K_{q2}}{X'_q - X_l}\right), \tag{8.39}$$

$$A_{3,7} = \frac{\partial g_{E'_d}}{\partial E'_{dc}} = \frac{-1}{T'_{q0}}\left((X_q - X'_q)K_{q1}\frac{\partial I_q}{\partial E'_{dc}}\right), \tag{8.40}$$

$$\boldsymbol{A}_{4,1} = \frac{\partial g_{E'_q}}{\partial \alpha} = \frac{1}{T'_{d0}}\left((X_d - X'_d)K_{d1}\frac{\partial I_d}{\partial \alpha}\right), \tag{8.41}$$

$$\boldsymbol{A}_{4,3} = \frac{\partial g_{E'_q}}{\partial E'_d} = \frac{1}{T'_{d0}}\left((X_d - X'_d)K_{d1}\frac{\partial I_d}{\partial E'_d}\right), \tag{8.42}$$

$$\boldsymbol{A}_{4,4} = \frac{\partial g_{E'_q}}{\partial E'_q} = \frac{-1}{T'_{d0}}\left(1 + (X_d - X'_d)\left(-K_{d1}\frac{\partial I_d}{\partial E'_q} + \frac{K_{d2}}{X'_d - X_l}\right)\right), \tag{8.43}$$

$$\boldsymbol{A}_{4,5} = \frac{\partial g_{E'_q}}{\partial \Psi_{1d}} = \frac{1}{T'_{d0}}\left((X_d - X'_d)\left(K_{d1}\frac{\partial I_d}{\partial \Psi_{1d}} + \frac{K_{d2}}{X'_d - X_l}\right)\right), \tag{8.44}$$

$$\boldsymbol{A}_{4,6} = \frac{\partial g_{E'_q}}{\partial \Psi_{2q}} = \frac{1}{T'_{d0}}\left((X_d - X'_d)K_{d1}\frac{\partial I_d}{\partial \Psi_{2q}}\right), \tag{8.45}$$

$$\boldsymbol{A}_{4,7} = \frac{\partial g_{E'_q}}{\partial E'_{dc}} = \frac{1}{T'_{d0}}\left((X_d - X'_d)K_{d1}\frac{\partial I_d}{\partial E'_{dc}}\right), \tag{8.46}$$

$$\boldsymbol{A}_{4,8} = \frac{\partial g_{E'_q}}{\partial V_r} = \frac{-K_a}{T'_{d0}}, \tag{8.47}$$

$$\boldsymbol{A}_{5,1} = \frac{\partial g_{\Psi_{1d}}}{\partial \alpha} = \frac{X'_d - X_l}{T''_{d0}}\frac{\partial I_d}{\partial \alpha}, \tag{8.48}$$

$$\boldsymbol{A}_{5,3} = \frac{\partial g_{\Psi_{1d}}}{\partial E'_d} = \frac{1}{T''_{d0}}\left((X'_d - X_l)\frac{\partial I_d}{\partial E'_d}\right), \tag{8.49}$$

$$\boldsymbol{A}_{5,4} = \frac{\partial g_{\Psi_{1d}}}{\partial E'_q} = \frac{1}{T''_{d0}}\left(1 + (X'_d - X_l)\frac{\partial I_d}{\partial E'_q}\right), \tag{8.50}$$

$$\boldsymbol{A}_{5,5} = \frac{\partial g_{\Psi_{1d}}}{\partial \Psi_{1d}} = \frac{1}{T''_{d0}}\left((X'_d - X_l)\frac{\partial I_d}{\partial \Psi_{1d}} - 1\right), \tag{8.51}$$

$$\boldsymbol{A}_{5,6} = \frac{\partial g_{\Psi_{1d}}}{\partial \Psi_{2q}} = \frac{1}{T''_{d0}}\left((X'_d - X_l)\frac{\partial I_d}{\partial \Psi_{2q}}\right), \tag{8.52}$$

$$\boldsymbol{A}_{5,7} = \frac{\partial g_{\Psi_{1d}}}{\partial E'_{dc}} = \frac{1}{T''_{d0}}\left((X'_d - X_l)\frac{\partial I_d}{\partial E'_{dc}}\right), \tag{8.53}$$

$$A_{6,1} = \frac{\partial g_{\Psi_{2q}}}{\partial \alpha} = \frac{X'_q - X_l}{T''_{q0}} \frac{\partial I_q}{\partial \alpha}, \tag{8.54}$$

$$A_{6,3} = \frac{\partial g_{\Psi_{2q}}}{\partial E'_d} = \frac{1}{T''_{q0}} \left((X'_q - X_l) \frac{\partial I_q}{\partial E'_d} - 1 \right), \tag{8.55}$$

$$A_{6,4} = \frac{\partial g_{\Psi_{2q}}}{\partial E'_q} = \frac{X'_q - X_l}{T''_{q0}} \frac{\partial I_q}{\partial E'_q}, \tag{8.56}$$

$$A_{6,5} = \frac{\partial g_{\Psi_{2q}}}{\partial \Psi_{1d}} = \frac{X'_q - X_l}{T''_{q0}} \frac{\partial I_q}{\partial \Psi_{1d}}, \tag{8.57}$$

$$A_{6,6} = \frac{\partial g_{\Psi_{2q}}}{\partial \Psi_{2q}} = \frac{1}{T''_{q0}} \left((X'_q - X_l) \frac{\partial I_q}{\partial \Psi_{2q}} - 1 \right), \tag{8.58}$$

$$A_{6,7} = \frac{\partial g_{\Psi_{2q}}}{\partial E'_{dc}} = \frac{X'_q - X_l}{T''_{q0}} \frac{\partial I_q}{\partial E'_{dc}}, \tag{8.59}$$

$$A_{7,1} = \frac{\partial g_{E'_{dc}}}{\partial \alpha} = \frac{X''_d - X''_q}{T_c} \frac{\partial I_q}{\partial \alpha}, \tag{8.60}$$

$$A_{7,3} = \frac{\partial g_{E'_{dc}}}{\partial E'_d} = \frac{X''_d - X''_q}{T_c} \frac{\partial I_q}{\partial E'_d}, \tag{8.61}$$

$$A_{7,4} = \frac{\partial g_{E'_{dc}}}{\partial E'_q} = \frac{X''_d - X''_q}{T_c} \frac{\partial I_q}{\partial E'_q}, \tag{8.62}$$

$$A_{7,5} = \frac{\partial g_{E'_{dc}}}{\partial \Psi_{1d}} = \frac{X''_d - X''_q}{T_c} \frac{\partial I_q}{\partial \Psi_{1d}}, \tag{8.63}$$

$$A_{7,6} = \frac{\partial g_{E'_{dc}}}{\partial \Psi_{2q}} = \frac{X''_d - X''_q}{T_c} \frac{\partial I_q}{\partial \Psi_{2q}}, \tag{8.64}$$

$$A_{7,7} = \frac{\partial g_{E'_{dc}}}{\partial E'_{dc}} = \frac{1}{T_c} \left(\frac{\partial I_q}{\partial E'_{dc}} (X''_d - X''_q) - 1 \right), \tag{8.65}$$

$$A_{8,8} = \frac{\partial g_{V_r}}{\partial V_r} = \frac{-1}{T_r}, \tag{8.66}$$

$$\boldsymbol{B}_{4,1} = \frac{\partial g_{E'_q}}{\partial V_{ss}} = \frac{K_a}{T'_{d0}}, \tag{8.67}$$

$$\boldsymbol{B}'_{1,2} = \frac{\partial g_{\alpha}}{\partial f} = -\omega_B, \tag{8.68}$$

$$\boldsymbol{B}'_{2,1} = \frac{\partial g_{\omega}}{\partial V} = \frac{-1}{2H}\left(\frac{\partial T_e}{\partial I_d}\frac{\partial I_d}{\partial V} + \frac{\partial T_e}{\partial I_q}\frac{\partial I_q}{\partial V}\right), \tag{8.69}$$

$$\boldsymbol{B}'_{3,1} = \frac{\partial g_{E'_d}}{\partial V} = \frac{-(X_q - X'_q)K_{q1}}{T'_{q0}}\frac{\partial I_q}{\partial V}, \tag{8.70}$$

$$\boldsymbol{B}'_{4,1} = \frac{\partial g_{E'_q}}{\partial V} = \frac{(X_d - X'_d)K_{d1}}{T'_{d0}}\frac{\partial I_d}{\partial V}, \tag{8.71}$$

$$\boldsymbol{B}'_{5,1} = \frac{\partial g_{\Psi_{1d}}}{\partial V} = \frac{X'_d - X_l}{T''_{d0}}\frac{\partial I_d}{\partial V}, \tag{8.72}$$

$$\boldsymbol{B}'_{6,1} = \frac{\partial g_{\Psi_{2q}}}{\partial V} = \frac{X'_q - X_l}{T''_{q0}}\frac{\partial I_q}{\partial V}, \tag{8.73}$$

$$\boldsymbol{B}'_{7,1} = \frac{\partial g_{E'_{dc}}}{\partial V} = \frac{X''_d - X''_q}{T_c}\frac{\partial I_q}{\partial V}, \tag{8.74}$$

$$\boldsymbol{B}'_{8,1} = \frac{\partial g_{V_r}}{\partial V} = \frac{1}{T_r}. \tag{8.75}$$

Discretizing (8.5) at a sampling period T_0 (T_0 is the sampling period of the dynamic state estimator) gives (see [77])

$$\boldsymbol{x}_{i(k+1)} = \boldsymbol{A}_i\boldsymbol{x}_{ik} + \boldsymbol{B}_i\boldsymbol{u}_{ik} + \boldsymbol{B}'_i\boldsymbol{u}'_{ik}, \tag{8.76}$$

$$\text{where } \boldsymbol{x}_{ik}=\Delta\boldsymbol{x}_{ci}(kT_0),\ \boldsymbol{u}_{ik}=\Delta\boldsymbol{u}_{ci}(kT_0),\ \boldsymbol{u}'_{ik}=\Delta\boldsymbol{u}'_{ci}(kT_0), \tag{8.77}$$

$$\boldsymbol{A}_i = e^{\boldsymbol{A}_{ci}T_0},\ \boldsymbol{B}_i = \boldsymbol{A}_{ci}^{-1}(\boldsymbol{A}_i - \boldsymbol{I})\boldsymbol{B}_{ci},\ \boldsymbol{B}'_i = \boldsymbol{A}_{ci}^{-1}(\boldsymbol{A}_i - \boldsymbol{I})\boldsymbol{B}'_{ci}. \tag{8.78}$$

Writing (8.76) in simplified form by dropping suffix i, we have

$$\boldsymbol{x}_{k+1} = \boldsymbol{A}\boldsymbol{x}_k + \boldsymbol{B}\boldsymbol{u}_k + \boldsymbol{B}'\boldsymbol{u}'_k. \tag{8.79}$$

Remark. The frequencies of the electromechanical modes of a machine lie in the range of 1.5–3.0 Hz [63]. As the ELQR controllers need to control and properly damp these modes, the minimum required sampling frequency for the controller is 6.0 Hz according to the Nyquist–Shannon sampling theorem (i.e., a maximum allowed sampling period of 0.17 s). This upper limit is also the threshold requirement for the sampling period. The lower limit is decided by the rate at which the dynamic states are provided to the controllers, which is given by T_0. As it is desired that the controllers operate at the fastest update rate possible, T_0 is also used as the sampling period for finding the discrete model and the control laws.

Remark. It should also be noted that if PMUs are used for dynamic estimation, then PMUs are required not only for DSE, but also for the ELQR control. The ELQR requires the dynamic state estimates provided by DSE and the phasor measurements provided by the PMU for the calculation of control gains, as shown in Fig. 8.1. Alternatively, if dynamic estimation is performed using CT/VT measurements (without using PMUs), then ELQR requires the dynamic state estimates and the measurements provided by CT/VT.

8.3 INTEGRATED ELQR CONTROL

The discrete equation in (8.79) has an extra term (corresponding to the pseudoinputs) as compared to the general discrete-time linear time-invariant system given by (7.1). It should be understood that the extra term $\boldsymbol{u}'_k$ cannot be absorbed in $\boldsymbol{u}_k$ as $\boldsymbol{u}'_k$ is an exogenous input while $\boldsymbol{u}_k$ is a normal control input. Thus the optimal control policy for (8.79) gets modified from traditional LQR to ELQR (Theorem 7.1), as explained in Chapter 7.

The terms $\boldsymbol{G}_k$ and $\boldsymbol{S}_k$ in Theorem 7.1 remain independent of the sequence of $\boldsymbol{u}'_k$, and hence they can be easily calculated if $\boldsymbol{A}$, $\boldsymbol{B}$, $\boldsymbol{B}'$, $\boldsymbol{Q}$, and $\boldsymbol{R}$ are known. On the other hand, the terms $\boldsymbol{G}'_k$ and $\boldsymbol{S}'_k$ require the knowledge of the sequence of $\boldsymbol{u}'_k$ for all the future and present samples, and hence they cannot be calculated for a practical power system as only the past and present values of the sequence of $\boldsymbol{u}'_k$ are available. Moreover, using offline values of the pseudoinputs (which are V and θ) it is found that $\boldsymbol{G}'_k$ makes a very small contribution in the control law given by Theorem 7.1. Thus, while implementing ELQR, $\boldsymbol{G}'_k$ is ignored and only the optimal gains $\boldsymbol{F}_k$ and $\boldsymbol{G}_k$ are calculated in real-time. Also, Theorem 7.1 requires that $\boldsymbol{u}'_k = 0 \; \forall \; k \geq N$. This condition can be taken into account if $N \to \infty$, that is, if no limit is imposed on the time within which the power system comes to a steady state. As $N \to \infty$ is the infinite-horizon case, the final decentralized control policy (using Theorem 7.1), after including the suffix i for the ith unit (whose equation is given by (8.76)), is written

$$\boldsymbol{u}_{ik} = -(\boldsymbol{F}_i \boldsymbol{x}_{ik} + \boldsymbol{G}_i \boldsymbol{u}'_{ik}), \; k \geq 0, \; \boldsymbol{Q}_i \geq 0, \; \boldsymbol{R}_i > 0, \tag{8.80}$$

$$\boldsymbol{F}_i = (\boldsymbol{R}_i + \boldsymbol{B}_i^T \boldsymbol{P}_i \boldsymbol{B}_i)^{-1} \boldsymbol{B}_i^T \boldsymbol{P}_i \boldsymbol{A}_i, \tag{8.81}$$

$$\boldsymbol{P}_i = \boldsymbol{Q}_i + \boldsymbol{A}_i^T [\boldsymbol{P}_i - \boldsymbol{P}_i \boldsymbol{B}_i (\boldsymbol{R}_i + \boldsymbol{B}_i^T \boldsymbol{P}_i \boldsymbol{B}_i)^{-1} \boldsymbol{B}_i^T \boldsymbol{P}_i] \boldsymbol{A}_i, \tag{8.82}$$

$$\boldsymbol{G}_i = \boldsymbol{F}_i (\boldsymbol{A}_i - \boldsymbol{P}_i^{-1} (\boldsymbol{P}_i - \boldsymbol{Q}_i))^{-1} \boldsymbol{B}'_i. \tag{8.83}$$

8.3.1 Damping control

The stable response of a power system requires that all the critical electromechanical modes in the system have damping ratios more than a certain percentage (typically more than 10%), as explained in Chapter 2. This can be achieved by ensuring that each unit provides a minimum damping to the intraplant mode it observes, and the collective damping efforts of all the units lead to damping of all the intraarea and interarea oscillations in the system. This constraint implies that the electromechanical poles (or modes) observed at a unit should lie within a conic section in the left half of the s-plane. In the z-plane, the conic section maps to a logarithmic spiral [123], and hence the discrete-domain poles should lie within the spiral. But confining the closed-loop poles within a logarithmic spiral is not practical; a practical alternative is to substitute the spiral with a disk and confine the closed-loop poles of the system within that disk. It is this technique that is used in this chapter for damping control.

Using Theorem 7.1 in Chapter 7 and Theorem 2 in [124] it can be shown that the decentralized control policy of ELQR for confining the closed-loop poles within a disk of radius r and center $(\beta, 0)$ remains the same as in (8.80)–(8.83) except that $\boldsymbol{A}_i$, $\boldsymbol{B}_i$, and $\boldsymbol{B}'_i$ are replaced by $(\boldsymbol{A}_i - \beta \boldsymbol{I})/r$, $\boldsymbol{B}_i/r$, and $\boldsymbol{B}'_i/r$, respectively. This technique requires a circle which coincides with the logarithmic spiral at the points where the electromechanical poles should lie. As electromechanical poles have high participation from the states of δ and ω, there is only one pair of electromechanical intraplant modes for a machine (as each machine has only one pair of δ and ω). Let the modal frequency of this intraplant mode be

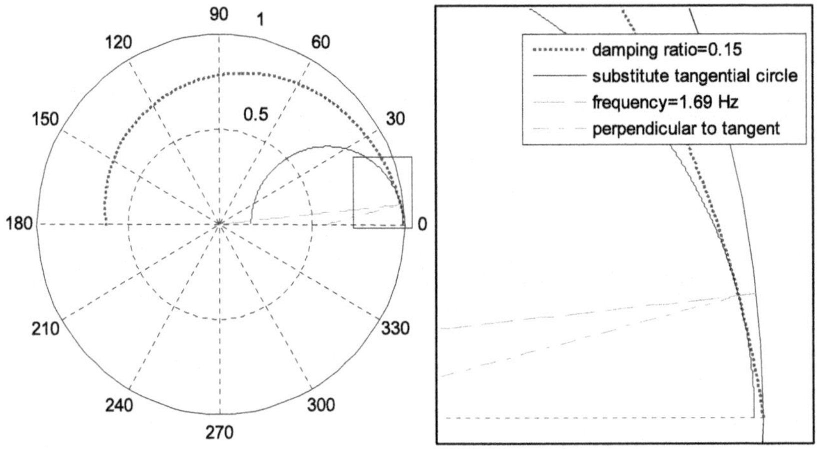

Figure 8.3 Circle substituting a logarithmic spiral.

f_m and let the minimum damping ratio to which this mode needs to be damped be d_{min}. Since it is desired that the substituting circle exactly coincides with the logarithmic spiral at the point corresponding to (f_m, d_{min}), the substituting circle should intersect the spiral at this point and it should also be inside the spiral. This can only happen when the circle is tangential to the spiral at this point from within the spiral. This substitution of the spiral with a circle can be better understood from Fig. 8.3.

In Fig. 8.3, the blue-dotted (black-dotted in print version) spiral corresponds to a constant damping ratio of $d_{min} = 0.15$ (only the upper half has been shown, the lower half will be its mirror image), the red-dashed (light gray-dashed in print version) line corresponds to a constant frequency of $f_m = 1.69$ Hz (this is the modal frequency of the intraplant mode of the 9th machine; all the calculations for this machine have been shown in the case study in Section 8.4) and the black-solid curve is the substituting circle. All the curves are inside the unity circle. The substituting circle should be tangent at the point where the constant frequency line intersects the constant damping ratio spiral. The black-solid curve denotes this tangential circle. For clarity, the right subfigure in Fig. 8.3 shows the magnified version of the region enclosed by the small rectangle in the left subfigure. This substituting circle ensures a damping ratio of more than or equal to d_{min} for all the poles of the machine, as the circle is completely inside the spiral, and the damping ratio of d_{min} is exactly ensured for the intraplant mode of modal frequency f_m as the circle will be tangential to the spiral at the point corresponding to (f_m, d_{min}). Thus, the parameters of this circle can be used in deriving the modified ELQR law for damping the intraplant modes. Using coordinate geometry, the parameters r and β for the circle for given f_m and d_{min} are found as follows:

$$\beta = R(\cos\theta_m - \frac{\sin\theta_m}{M}),\quad r = \sqrt{R^2 + \beta^2 - 2\beta R\cos\theta_m},$$
$$\text{where } R = e^{-d_m\theta_m},\ d_m = \cot(\cos^{-1} d_{min}),\ \theta_m = 2\pi f_m T_0,$$
$$\text{and } M = (\sin\theta_m + d_m\cos\theta_m)/(\cos\theta_m - d_m\sin\theta_m). \tag{8.84}$$

8.4 CASE STUDY

8.4.1 System description

The test system remains the same as the one used in Chapter 4 (Fig. 4.3). As explained in Chapter 2, this system has four interarea modes in the range 0.2–1.0 Hz, three of which are poorly damped with damping ratios less than 10%.

Each machine in the system is assumed to be equipped with an excitation system controller, a PSS, a dynamic state estimator, and an ELQR controller. PSS control is used only for comparison, that is, in one case only ELQR is working and in the second case only PSS is working. They are not working together in any case. The control case when only PSS is working has been termed "PSS control," while the case when only ELQR is working has been termed "ELQR control".

The matrices $\boldsymbol{Q}$ and $\boldsymbol{R}$ are positive semidefinite and positive definite matrices, respectively, and their values depend on how costs/penalties are assigned to the deviations of the states and inputs, respectively, from their steady-state values. In the case study it is desired that the sum of the squares of deviations for all the states and all the inputs for a machine is minimized for all the time samples, so that all the state and input deviations get uniform penalties in the control law. Hence, the state cost for the ith machine is taken as $\sum_{k=0}^{N-1}(\sum_{j=1}^{7}|x_{ijk}|^2)$ for the seven states of the ith machine. Since $\sum_{j=1}^{7}|x_{ijk}|^2 = \boldsymbol{x}_{ik}^T\boldsymbol{x}_{ik} = \boldsymbol{x}_{ik}^T\boldsymbol{I}_7\boldsymbol{x}_{ik}$, $\boldsymbol{Q} = \boldsymbol{I}_7$ for each machine ($\boldsymbol{I}_7$ is an identity matrix of order 7). Similarly, control cost is $\sum_{k=0}^{N-1}(|u_{ik}|^2)$ (as there is only one control input); and since $|u_{ik}|^2 = u_{ik}u_{ik} = u_{ik}.1.u_{ik}$, $\boldsymbol{R} = 1$.

The state estimator provides estimates every 8.33 ms, while the state matrices and the control gains of the ELQR are updated every second. As an example, the complete calculation process for finding the control gains for one of the machines (the 9th machine) at $t = 0$ has been shown as follows. The calculation process remains the same for the rest of the machines in the system.

The constant parameters for the 9th machine, using the data for the 68-bus system from Appendix A, are
$X_l = 0.0298$ p.u., $R_a = 0$ p.u., $X_d = 0.2106$ p.u., $X_d' = 0.057$ p.u., $X_d'' = 0.045$ p.u., $X_q = 0.205$ p.u., $X_q' = 0.05$ p.u., $X_q'' = 0.045$ p.u., $T_{d0}' = 4.79$ s, $T_{d0}'' = 0.05$ s, $T_{q0}' = 1.96$ s, $T_{q0}'' = 0.035$ s, $D = 14$ p.u., $H = 34.5$ s, $\omega_B = 376.99$ rad/s, $K_a = 10$ p.u., $T_r = 0.01$ s.

The values of the states and algebraic variables for the 9th machine at $t = 0$, found using DSE, are
$\alpha = 0.950$ rad, $\omega = 1$ p.u., $E_d' = -0.630$ p.u., $E_q' = 0.978$ p.u., $\Psi_{1d} = 0.796$ p.u., $\Psi_{2q} = 0.713$ p.u., $E_{dc}' = 0$ p.u., $V_r = 1.025$ p.u., $E_{fd} = 2.005$ p.u., $I_d = -6.687$ p.u., $I_q = 4.067$ p.u., $V_d = -0.834$ p.u., $V_q = 0.596$ p.u., $T_e = 8$ p.u.

As $X_d'' = X_q''$ for all the machines of the 68-bus system, E_{dc}' remains constant (equal to zero) and can be eliminated from the DAEs and the linearized equations. Thus, there are effectively seven dynamic states for each machine in the system. The following system matrices are found for the 9th machine after substituting the above values of the parameters and states into the expressions for $\boldsymbol{A}_c$, $\boldsymbol{B}_c$, and $\boldsymbol{B}_c'$ in Section 8.2.1 and eliminating expressions corresponding to E_{dc}':

$$\boldsymbol{B}_c = \begin{bmatrix} 0 \\ 0 \\ 0 \\ 2.088 \\ 0 \\ 0 \\ 0 \end{bmatrix}, \quad \boldsymbol{B}_c' = \begin{bmatrix} 0 & -376.9 \\ -0.113 & 0 \\ -1.076 & 0 \\ 0.232 & 0 \\ 7.034 & 0 \\ 10.43 & 0 \\ 100 & 0 \end{bmatrix}, \tag{8.85}$$

$$\boldsymbol{A}_c = \begin{bmatrix} 0 & 376.9 & 0 & 0 & 0 & 0 & 0 \\ -0.347 & -0.203 & -0.145 & -0.150 & -0.118 & 0.048 & 0 \\ -0.789 & 0 & -2.474 & 0 & 0 & -0.642 & 0 \\ -0.332 & 0 & 0 & -0.951 & 0.344 & 0 & -2.088 \\ -10.08 & 0 & 0 & 13.24 & -25.33 & 0 & 0 \\ 7.649 & 0 & -18.92 & 0 & 0 & -31.75 & 0 \\ 0 & 0 & 0 & 0 & 0 & 0 & -100 \end{bmatrix}. \tag{8.86}$$

The discrete forms of the above matrices, using (7.10) ($T_0 = 0.00833$ s), are

$$\boldsymbol{B} = \begin{bmatrix} 0 \\ 0 \\ 0 \\ 0.017 \\ 0.001 \\ 0 \\ 0 \end{bmatrix}, \boldsymbol{B}' = \begin{bmatrix} -0.002 & -3.137 \\ -0.001 & 0.004 \\ -0.009 & 0.010 \\ -0.004 & 0.004 \\ 0.053 & 0.123 \\ 0.077 & -0.092 \\ 0.565 & 0 \end{bmatrix}, \tag{8.87}$$

$$\boldsymbol{A} = \begin{bmatrix} 0.996 & 3.134 & -0.002 & -0.002 & -0.001 & 0.001 & 0 \\ -0.003 & 0.994 & -0.001 & -0.001 & -0.001 & 0.000 & 0 \\ -0.007 & -0.010 & 0.980 & 0 & 0 & -0.005 & 0 \\ -0.003 & -0.004 & 0 & 0.992 & 0.003 & 0 & -0.012 \\ -0.076 & -0.123 & 0 & 0.099 & 0.810 & 0 & -0.001 \\ 0.056 & 0.092 & -0.137 & 0 & 0 & 0.768 & 0 \\ 0 & 0 & 0 & 0 & 0 & 0 & 0.435 \end{bmatrix}. \tag{8.88}$$

Using the above value of $\boldsymbol{A}$, the intraplant modes are found as -0.895 ± 10.617. The modal frequency for this pair of modes is $f_m = 10.617/(2\pi) = 1.69$ Hz. Finally, using Eqs. (8.80)–(8.83), after replacing $\boldsymbol{A}$, $\boldsymbol{B}$, and $\boldsymbol{B}'$ with $(\boldsymbol{A} - \beta \boldsymbol{I})/r$, $\boldsymbol{B}/r$, and $\boldsymbol{B}'/r$, respectively (as explained in Section 8.3.1), taking $r = 0.411$, $\beta = 0.581$ (which are found using (8.84), after taking $f_m = 1.69$ Hz and $d_{min} = 0.15$) and taking $\boldsymbol{Q}$ and $\boldsymbol{R}$ as identity matrices, the gain matrices $\boldsymbol{F}$ and $\boldsymbol{G}$ for the 9th machine are found to be

$$\boldsymbol{F} = \begin{bmatrix} 2.448 & -20.524 & 2.143 & 3.378 & 0.288 & -0.110 & -0.070 \end{bmatrix}, \tag{8.89}$$

$$\boldsymbol{G} = \begin{bmatrix} 0.098 & 12.316 \end{bmatrix}. \tag{8.90}$$

At each unit, a washout filter with time constant of 10 s is also applied to the ELQR output signal. This ensures that the steady-state output of the ELQR is zero to allow operation of the system at off-nominal frequency. The output signal from the ELQR can also get unbounded transiently during contingencies; therefore its output is limited just like a PSS, with $|V_{ss}| < 0.01$ p.u. Although the parameters for the excitation system, PSS,

Table 8.1 Modal analysis for the four interarea modes

	Without control	PSS control	ELQR control	WADC control
Mode-1 frequency (Hz)	0.39	0.44	0.31	0.44
Mode-1 damping ratio (%)	2.1	14.8	21.9	20.6
Mode-2 frequency (Hz)	0.52	0.54	0.47	0.52
Mode-2 damping ratio (%)	2.7	7.1	10.9	17.2
Mode-3 frequency (Hz)	0.60	0.63	0.54	0.66
Mode-3 damping ratio (%)	1.9	7.0	12.1	11.4
Mode-4 frequency (Hz)	0.79	0.81	0.76	0.80
Mode-4 damping ratio (%)	4.8	7.0	10.5	12.8

and the washout filter can be tuned individually for each machine in the system, in the case study standard parameters have been used, given in Appendix A. This is done so that the performance of the ELQR methodology is evaluated in a standard framework. The system is simulated in MATLAB-Simulink. Level-2 S-functions are used for dynamic updating of state matrices and control gains.

8.4.2 Control performance

In the simulation, the system starts from steady state, and then a balanced three-phase fault is applied in one of the tie lines between buses 53–54 followed by immediate outage of this tie line. Fig. 8.4 shows the plots of the relative rotor speed between machines 13–16 and the power flow in interarea tie lines between buses 60–61 for two cases. In the first case each machine is controlled using PSS control, while in the second case each machine is controlled using ELQR control. Table 8.1 shows the modal frequencies and damping ratios for the four poorly damped interarea modes. It can be observed that although the modal frequencies for the ELQR case decrease as compared to the case without control, this decrease is strongly compensated by the increase in damping ratios of these modes, and all the modes are damped within damping ratios of 10% or more. A similar improvement in damping performance is not observed for the case of PSS control. Thus, Fig. 8.4 and Table 8.1 show that the control and damping performance of ELQR control is significantly better than that of PSS control.

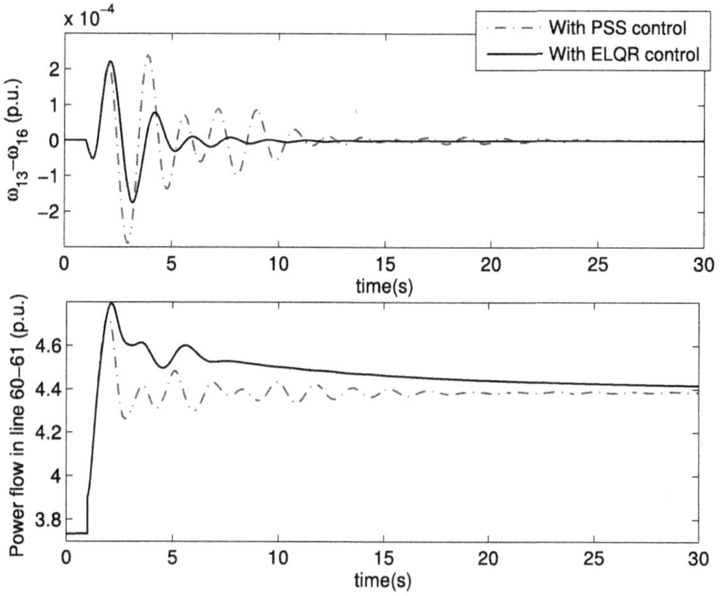

Figure 8.4 Dynamic performance of PSS control vs. ELQR control.

8.4.3 Robustness to different operating conditions

As the state matrices and control gains are updated every second and get adapted to the current system conditions, the control remains valid for any operating point. The power flow in line 60–61 has been shown for three operating cases. The total power flow from the area NETS to the area NYPS is varied in the three cases; it is 700 MW for the first case (Fig. 8.4, second plot), 100 MW for the second case (Fig. 8.5, first plot), and 900 MW for the third case (Fig. 8.5, second plot). It can be observed that ELQR control remains robust under varying operating conditions.

8.4.4 Control efforts and state costs

Fig. 8.6 shows the 13th unit's control signal (which is V_{ss13}) for PSS control and ELQR control. It can be seen that ELQR has a lower magnitude for the control signal than PSS control. But, the value of the control signal for a unit (or even for all the units, if each unit is considered separately) is inconclusive. Better metrics for evaluating the quality of a control method are control efforts and state costs associated with that method. Table 8.2 presents a comparison of total cost given by $\sum_{i=1}^{m}\sum_{k=0}^{N-1}\{\boldsymbol{u}_{ik}^{T}\boldsymbol{u}_{ik}+\boldsymbol{x}_{ik}^{T}\boldsymbol{x}_{ik}\}$, which is the sum of control efforts (or the

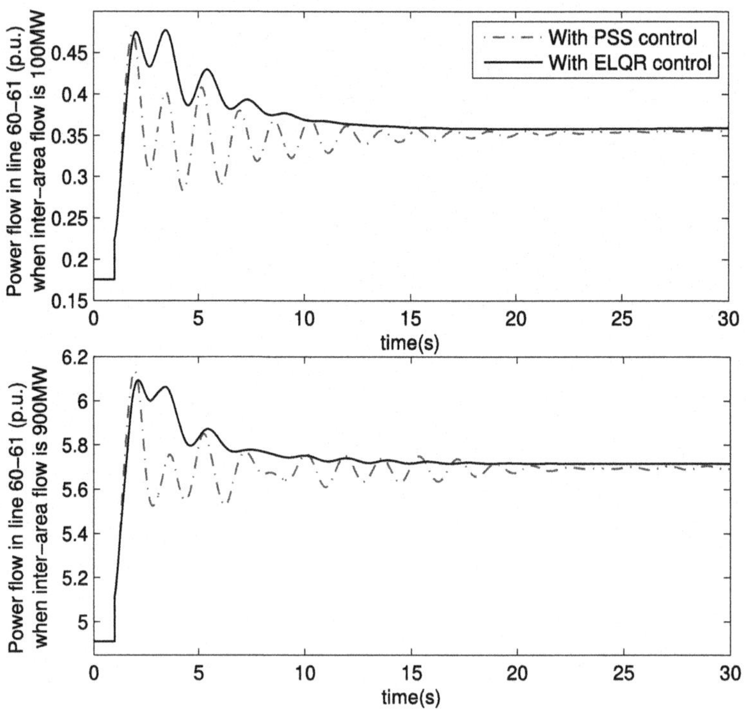

Figure 8.5 Dynamic performance for different operating conditions.

control cost $= \sum_{i=1}^{m} \sum_{k=0}^{N-1} \{\boldsymbol{u}_{ik}^T \boldsymbol{u}_{ik}\})$ and state costs $(= \sum_{i=1}^{m} \sum_{k=0}^{N-1} \{\boldsymbol{x}_{ik}^T \boldsymbol{x}_{ik}\})$. Three more operating cases are shown in Table 8.2, in which the faulted tie line has been changed. It can be observed that although the control costs for PSS control and ELQR control are similar, the state costs are reduced by an average of 28.2% and the total costs are reduced by an average of 24.3% for ELQR control as compared to PSS control.

8.4.5 Comparison with centralized wide area-based control

A wide-area damping control (WADC)-based control given in [20] has also been used for comparison with the DSE-ELQR scheme. In this scheme, wide-area signals which have high observability of the intraarea and interarea electromechanical modes are used to control excitation systems of several machines which have high controllability of those modes. The control signal V_{ss} is used for this purpose, which is the same as the control signal used by a PSS or ELQR. The design of the centralized WADC controller is made using the linearized and reduced model of the whole system

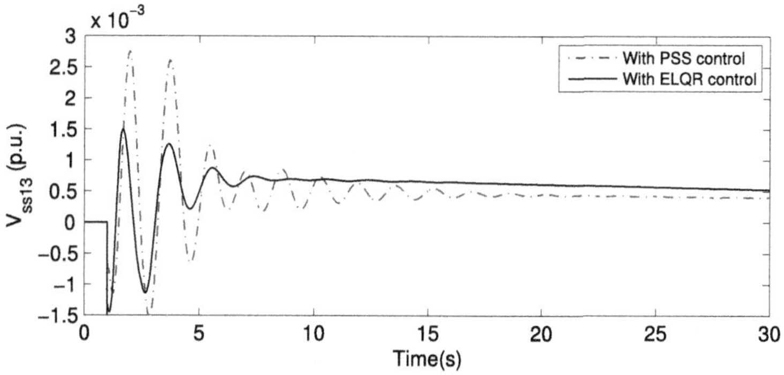

Figure 8.6 Comparison of the values of the control signal for unit 13.

Table 8.2 Comparison of total costs

Operating condition: interarea power flow and faulted tie line	Total cost (state cost + control cost) for PSS control (p.u.)	Total cost (state cost+ control cost) for ELQR control (p.u.)
700 MW, 53–54	2.244 (1.908 + 0.336)	1.679 (1.391 + 0.288)
100 MW, 53–54	0.360 (0.312 + 0.048)	0.214 (0.165 + 0.049)
900 MW, 53–54	3.732 (3.228 + 0.504)	2.951 (2.491 + 0.460)
700 MW, 27–53	0.212 (0.192 + 0.020)	0.184 (0.155 + 0.029)
100 MW, 27–53	0.148 (0.132 + 0.016)	0.102 (0.084 + 0.018)
900 MW, 27–53	0.259 (0.228 + 0.031)	0.221 (0.191 + 0.030)

and its tuning is based on mixed H_2/H_∞ optimization with pole placement constraints, as detailed in [20]. Seven power flow signals are used as output measurements: P_{13-17}, P_{16-18}, P_{3-62}, P_{9-29}, P_{15-42}, P_{10-31}, and P_{20-19}. Each one of these signals has the highest observability of one or more intraarea/interarea modes of the system. Using these signals and the designed controller, control inputs (V_{ss}) are sent to the excitation system of each machine in the system. It is possible to select only some machines for wide-area control, but as all the machines are controlled in decentralized ELQR, in centralized WADC also all of the 16 machines have been selected for uniformity in comparison. Comparisons of time-domain simulation, modal response, and control/state costs are shown in Fig. 8.7, Table 8.1, and Table 8.3, respectively.

It can be observed from Fig. 8.7 and Table 8.1 that the damping performance of ELQR control and WADC are comparable, and WADC gives better damping ratios to the second and fourth interarea modes, while

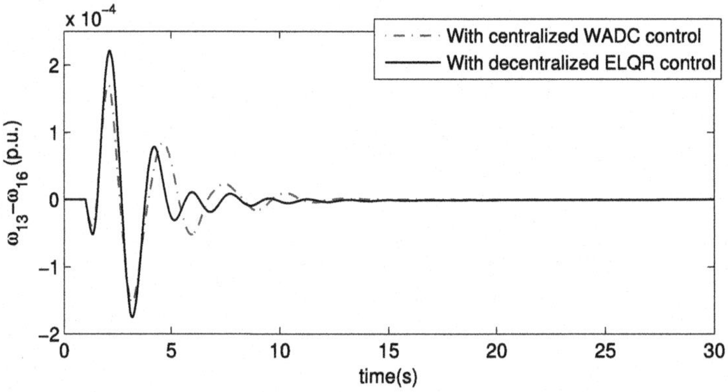

Figure 8.7 Oscillation damping comparison for WADC and ELQR.

Table 8.3 Comparison of total costs for WADC vs. ELQR

Operating condition: interarea power flow and faulted tie line	Total cost (state cost+ control cost) for WADC control (p.u.)	Total cost (state cost+ control cost) for ELQR control (p.u.)
700 MW, 53–54	2.580 (2.424 + 0.156)	1.679 (1.391 + 0.288)
100 MW, 53–54	0.372 (0.348 + 0.024)	0.214 (0.165 + 0.049)
900 MW, 53–54	4.440 (4.164 + 0.276)	2.951 (2.491 + 0.460)
700 MW, 27–53	0.276 (0.264 + 0.012)	0.184 (0.155 + 0.029)
100 MW, 27–53	0.184 (0.180 + 0.004)	0.102 (0.084 + 0.018)
900 MW, 27–53	0.312 (0.298 + 0.014)	0.221 (0.191 + 0.030)

ELQR control gives better damping to the first and third modes. The costs (as shown in Table 8.3) are not as uniformly distributed between state costs and control efforts as in the case of ELQR control, and thus the total costs are higher for WADC than ELQR control. This is expected as mixed H_2/H_∞ optimization is not as optimal as ELQR control as far as net quadratic costs are concerned. Thus, it can be concluded that the DSE-ELQR scheme performs equally well as (or even better than) an established wide area-based centralized damping control method. Considering the fact that WADC requires information of the whole system for controller design and requires a fast and reliable communication network for transmission of measurements and control signals, decentralized ELQR is a better choice than centralized WADC.

Table 8.4 Comparison of total cost with and without noise/bad data

Condition: interarea flow, faulted tie line	Total cost with noise (state cost + control cost) for ELQR control (p.u.)	Total cost without noise (state cost + control cost) for ELQR control (p.u.)
700 MW, 53–54	1.679 (1.391 + 0.288)	1.675 (1.390 + 0.285)
100 MW, 53–54	0.214 (0.165 + 0.049)	0.212 (0.164 + 0.048)
900 MW, 53–54	2.951 (2.491 + 0.460)	2.947 (2.490 + 0.457)
700 MW, 27–53	0.184 (0.155 + 0.029)	0.183 (0.154 + 0.029)
100 MW, 27–53	0.102 (0.084 + 0.018)	0.102 (0.084 + 0.018)
900 MW, 27–53	0.221 (0.191 + 0.030)	0.220 (0.190 + 0.030)

8.4.6 Effect of noise/bad data on control performance

ELQR control is affected by noise and bad data in the measurements, but the effect is too small to make any impact on the control performance. All the aforementioned results of the case study have been obtained considering noise and bad data in the measurements. For comparison, results have also been obtained without considering any noise or bad data in the measurements, and the costs are shown in Table 8.4.

It can be observed from Table 8.4 that the results of the case study remain almost the same, with or without noise in the measurements, and the state costs differ by an average of 0.2% and control costs differ by an average of 0.5%. The first reason for such a small change is that a majority of the contribution in the ELQR control output comes from the seven state estimates from which noise and bad data have been filtered out. Secondly, the level of noise in the measurements is very small: the standard deviation of the noise in magnitude measurements is 0.1% of true values and in phase measurements it is 0.1 mrad, for both voltage and current signals. These noise levels are as per IEC 60044/IEEE C57.13 standards for CTs and potential transformers (PTs) and IEEE C37.118.1-2011 standard for PMUs, as explained in Chapter 4. Such a small level of noise implies that the measurements deviate very little from their true values. Lastly, bad data in the measurements are detected, removed, and replaced with the latest correct data using the bad-data detection algorithm given in Chapter 4. Thus, noise and bad data have a negligible impact on the ELQR control performance.

8.4.7 Computational feasibility

The complete simulation of the power system, along with the dynamic estimators and the ELQR controllers at each of the 16 machines, runs in real-time. In the case study, a 30-s simulation takes an average running time of 5.5 s on a personal computer with Intel Core 2 Duo, 2.0 GHz CPU and 2 GB RAM. Hence the computational requirements for one machine can be easily met for both DSE and ELQR control.

8.5 SUMMARY

A control scheme has been presented for the decentralized control of power system dynamics. The scheme utilizes DSE using local PMU or CT/VT measurements and machine parameters and employs the concept of pseudoinputs for decentralization. The method is based on ELQR and adapts in real-time to varying operating conditions of the system. The method is also computationally feasible and easily implementable. The main features of this control architecture are:

1. Besides being optimal, the control is completely decentralized and only local measurements and machine parameters are needed, and hence communication requirements are minimized.
2. Computational requirements are less intensive, so they can be easily met by a personal computer.
3. Existing PMU in each decentralized location is adequate; no extra investment is required. If PMU is not available, then CT/VT are sufficient.
4. The control law remains valid for any operating condition, and the control gains are updated in real-time. This indirectly renders the control scheme adaptive to the current operating point.
5. The scheme can be seamlessly integrated with the control devices which are already present, such as PSSs and FACTS controllers.

In summary, it has been shown that the integrated scheme of DSE and ELQR can be utilized for dynamic estimation and control of small-signal dynamics of power systems in a decentralized manner, and it has several advantages over other linear control methods.

CHAPTER 9

Decentralized Nonlinear Control Using DSE & Normal Forms

While the previous chapter presented a decentralized scheme for linear control integrated with decentralized dynamic state estimation (DSE), this chapter presents a decentralized nonlinear control scheme, so that the method can ensure both transient and small signal stability of power systems. Normal form of power system dynamics have been derived using the theory discussed in Chapter 7. The developed normal form is further integrated with the theory of decentralized DSE for deriving a nonlinear control law for ensuring transient stability of power systems. As the employed control and estimation schemes only need local measurements, the method remains completely decentralized. Asymptotic stability of the controlled system and a comparison with existing methods of nonlinear control are also presented.

The majority of the control actions which are taken after a disturbance in a power system use linear control theory. This requires linearization of power system dynamics at a particular equilibrium point or at a finite set of equilibrium points [62,7]. Power systems are nonlinear in nature and a large disturbance (and sometimes even a small one [6]) can alter the operating point of the system quite significantly from equilibrium condition(s) [125]. As linearization is applicable only in the vicinity of equilibrium condition(s), application of linear control methods can prove to be ineffective at the altered condition and can possibly have an adverse effect on system stability [6,125]. A logical solution to this problem is to use nonlinear control methods which remain valid for any operating condition and not only for conditions which are close to equilibrium.

As explained in Chapter 7, normal form-based nonlinear control methods have established design steps and are also partially or fully linear. Hence, this chapter focuses on methods based on normal forms for nonlinear control of power systems.

Since the first half of the 1990s, an in-depth exploration of nonlinear control methods based on normal forms has been conducted for power systems, but almost all of these methods rely on model simplification and

Dynamic Estimation and Control of Power Systems
https://doi.org/10.1016/B978-0-12-814005-5.00020-0

approximations [47–52]. Specifically, some shortcomings which need to be addressed are the following.

1) A classical model is used to derive the normal form of power systems. The classical model is a reduced-order representation of a synchronous machine, and the transient dynamics of the system are incorrectly reflected in this model. Thus, using a control law which has been derived using this model can have unexpected or unwanted effects on the stability of the system.
2) The final control expression which is obtained is a function of unmeasurable states of the power system, such as rotor angle and transient flux. Approximations are made in order to represent this control expression as a function of measurable quantities, such as stator current and stator terminal voltage. As these measurements are acquired from instrument transformers and phasor measurement units (PMUs), noise, harmonics, and bad data are present in them [101–104]. Thus, the control expression can become grossly erroneous if it is approximated using these measurements, thereby impacting the stability of the system in a negative manner.

The above shortcomings can be addressed by the following two steps:

1) using a detailed subtransient model for the derivation of a normal form-based control law for power systems – this is the recommended model for transient stability analysis as per IEEE Std 1110-2002 [126] (also see [127]) and adequately models transient dynamics in power systems; and
2) using dynamic state estimates instead of PMU or CT/VT measurements for implementation of the derived control.

These two steps are further explored in the rest of the chapter. Sections 9.1–9.3 develop the normal form-based theory for power systems using a detailed model for machines. Section 9.4 explains the application of dynamic estimation in the developed nonlinear control. Section 9.5 presents simulations to demonstrate the practical applicability and implementability of the method using the 68-bus test system, and Section 9.6 summarizes the chapter.

9.1 NORMAL FORM OF POWER SYSTEM DYNAMICS

The subtransient model of machines with four rotor coils in each machine, known as Model 2.2 [126], has been used to study power system dynamics and to derive their normal form. A model of a static IEEE-ST1A excitation system is included with the model of each machine. Also, all the loads in the system are assumed to be modeled as constant impedance loads. The dynamic equations for this model are given in Chapter 2 and are summarized as follows (where i refers to the system's ith machine, $1 \leq i \leq M$). Also, ω is in rad/s for this chapter. We have

$$
\begin{aligned}
&\Delta\dot{\delta}^i = (\omega^i - \omega_b) = \Delta\omega^i \quad (\text{where, } \Delta\delta^i = \delta^i - \delta_0^i),\\
&\Delta\dot{\omega}^i = \frac{\omega_b}{2H^i}(T_m^i - T_e^i) - \frac{D^i}{2H^i}\Delta\omega^i,\\
&\dot{E}_d'^i = \frac{1}{T_{q0}'^i}[-E_d'^i - (X_q^i - X_q'^i)[K_{q1}^i I_q^i + K_{q2}^i \frac{\Psi_{2q}^i + E_d'^i}{X_q'^i - X_l^i}]],\\
&\dot{E}_q'^i = \frac{E_{fd}^i - E_q'^i + (X_d^i - X_d'^i)[K_{d1}^i I_d^i + K_{d2}^i \frac{\Psi_{1d}^i - E_q'^i}{X_d'^i - X_l^i}]}{T_{d0}'^i},\\
&\dot{\Psi}_{1d}^i = \frac{1}{T_{d0}''^i}[E_q'^i + (X_d'^i - X_l^i)I_d^i - \Psi_{1d}^i],\\
&\dot{\Psi}_{2q}^i = \frac{1}{T_{q0}''^i}[-E_d'^i + (X_q'^i - X_l^i)I_q^i - \Psi_{2q}^i],\\
&\dot{E}_{dc}'^i = \frac{1}{T_c^i}[(X_d''^i - X_q''^i)I_q^i - E_{dc}'^i], \qquad (9.1)\\
&\dot{V}_r^i = \frac{1}{T_r^i}[V^i - V_r^i], \text{ where}\\
&E_{fd}^i = K_a^i[V_{ss}^i + V_{ref}^i - V_r^i], \; E_{fdmin}^i \leq E_{fd}^i \leq E_{fdmax}^i,\\
&\begin{bmatrix} I_d^i \\ I_q^i \end{bmatrix} = \begin{bmatrix} R_s^i & X_q''^i \\ -X_d''^i & R_s^i \end{bmatrix}^{-1} \begin{bmatrix} E_d'^i K_{q1}^i - \Psi_{2q}^i K_{q2}^i - V_d^i \\ E_q'^i K_{d1}^i + \Psi_{1d}^i K_{d2}^i - V_q^i \end{bmatrix},\\
&T_e^i = \frac{\omega_b}{\omega^i} P_G^i, \; P_G^i = V_d^i I_d^i + V_q^i I_q^i, \; Q_G^i = V_d^i I_q^i - V_q^i I_d^i,\\
&V_d^i = -V^i \sin(\delta^i - \theta^i), \quad V_q^i = V^i \cos(\delta^i - \theta^i),\\
&I_g^i = \frac{[E_q'^i K_{d1}^i + \Psi_{1d}^i K_{d2}^i + j\{E_d'^i K_{q1}^i - \Psi_{2q}^i K_{q2}^i - E_{dc}'^i\}]e^{j\delta^i}}{R_s^i + jX_d''^i}.
\end{aligned}
$$

Using the facts that $R_s << X_q''$, that $R_s << X_d''$, and that R_s is normally taken as 0 p.u. in power system studies ([62], [2]), the above expression for

I_d^i and I_q^i is simplified as follows:

$$
\begin{aligned}
I_d^i &= (V_q^i - E_q'^i K_{d1}^i - \Psi_{1d}^i K_{d2}^i)/X_d''^i, \\
I_q^i &= (E_d'^i K_{q1}^i - \Psi_{2q}^i K_{q2}^i - V_d^i)/X_q''^i.
\end{aligned}
\tag{9.2}
$$

The bus voltages, (V^i, θ^i), $1 \leq i \leq N$ (the total number of buses is N), are given by the following load flow equations (as explained in Chapter 2):

$$
\begin{aligned}
P^i &= \sum_{j=1}^{N} V^i V^j [G^{ij} \cos(\theta^i - \theta^j) + B^{ij} \sin(\theta^i - \theta^j)], \\
Q^i &= \sum_{j=1}^{N} V^i V^j [G^{ij} \sin(\theta^i - \theta^j) - B^{ij} \cos(\theta^i - \theta^j)].
\end{aligned}
\tag{9.3}
$$

In order to apply nonlinear control theory to power systems, the differential and algebraic equations (DAEs) of the power system given by (9.1) and (9.3) need to be mathematically shown equivalent to the affine ordinary differential equations (ODEs) given by (7.78). This equivalence can be shown for a well-defined network configuration. Here, a network configuration is considered to be well defined if the matrix corresponding to the equivalent admittance of the network exists and is nonsingular. One example in which the network configuration is not well defined is during a fault in which the line admittance of one or more lines becomes infinite, and hence the line admittance matrix is undefined and the equivalence fails to hold for the duration of the fault. For a well-defined network configuration, one way to show the equivalence of these DAEs and ODEs is by representing each machine as a current source, I_g^i (defined in (9.1)), behind a constant admittance, $1/(R_s^i + jX_d''^i)$. It should be noted that I_g^i is a function of only the states of the ith machine and not of any algebraic variables.

The network equations in (9.3) can be written as follows using basic relations between voltages and current injections:

$$
\boldsymbol{V} = \boldsymbol{Z}_A \boldsymbol{I}_g, \ \boldsymbol{Z}_A = (\boldsymbol{Y}_A)^{-1}, \ \boldsymbol{Y}_A = \boldsymbol{Y}_N + \boldsymbol{Y}_G + \boldsymbol{Y}_L. \tag{9.4}
$$

As mentioned in Chapter 2, here $\boldsymbol{V}$ is the column vector of bus voltages; $\boldsymbol{I}_g$ is the column vector of current injections, with the ith element equal to I_g^i if i is a machine bus, else it is equal to zero; $\boldsymbol{Y}_N$ is the network admittance matrix; $\boldsymbol{Y}_G$ is the diagonal matrix of machine admittances, with the ith diagonal element equal to $1/(R_s^i + jX_d''^i)$ if i is a machine bus, else it is equal to zero; similarly, $\boldsymbol{Y}_L$ is the diagonal matrix of load admittances; and $\boldsymbol{Y}_A$ and $\boldsymbol{Z}_A$ are augmented matrices of admittance and impedance, respectively. It should be noted that $\boldsymbol{Y}_L$ will change with any change in load; hence, $\boldsymbol{Y}_L$ is a time-varying quantity, unless the loads are constant impedance loads. Thus, modeling loads as constant impedance loads is required to represent power system DAEs as ODEs given by (7.78). Also, $\boldsymbol{Y}_G$ and $\boldsymbol{Y}_N$ can always be found for a given well-defined network [62,63]. Thus, as it is assumed that the network configuration is well defined, $\boldsymbol{Y}_A$ and $\boldsymbol{Z}_A$ exist and are nonsingular.

The differential equations in (9.1) can be written as follows:

$$\dot{\boldsymbol{x}}^i = \boldsymbol{F}^i(\boldsymbol{x}^i, V_g^i) + \sum_{i=1}^{M} \boldsymbol{g}_i^i(\boldsymbol{x}^i)u_i, \tag{9.5}$$

where $\boldsymbol{x}^i$ is column vector of the ith machine's states, V_g^i is the machine's terminal bus voltage and is equal to $V^i e^{j\theta^i}$, $\boldsymbol{F}^i$ is the column vector of differential functions in (9.1), $\boldsymbol{g}_i^i(\boldsymbol{x}^i) = [0\ 0\ 0\ \frac{K_a^i}{T_{d0}^{\prime i}}\ 0\ 0\ 0\ 0]^T$, and $u_i = V_{ss}^i$. Consolidating (9.5) for $i = 1, 2, \ldots, M$ gives the following equation:

$$\dot{\boldsymbol{x}} = \boldsymbol{F}(\boldsymbol{x}, \boldsymbol{V}_g) + \sum_{i=1}^{M} \boldsymbol{g}_i(\boldsymbol{x})u_i. \tag{9.6}$$

Here $\boldsymbol{x} = [\boldsymbol{x}^{1^T} \boldsymbol{x}^{2^T} \ldots \boldsymbol{x}^{M^T}]^T$, $\boldsymbol{F} = [\boldsymbol{F}^{1^T} \boldsymbol{F}^{2^T} \ldots \boldsymbol{F}^{M^T}]^T$, $\boldsymbol{V}_g = [V_g^1\ V_g^2 \ldots V_g^M]^T$, $\boldsymbol{g}_i(\boldsymbol{x}) = [\boldsymbol{g}_i^{1^T} \boldsymbol{g}_i^{2^T} \ldots \boldsymbol{g}_i^{M^T}]^T$, where $\boldsymbol{g}_i^i$ is defined as before and $\boldsymbol{g}_i^j = [0\ 0\ 0\ 0\ 0\ 0\ 0\ 0]^T\ \forall j \neq i$. As $\boldsymbol{V}_g$ is a subset of $\boldsymbol{V}$ ($\boldsymbol{V}_g$ only constitutes voltages of machine buses, while $\boldsymbol{V}$ constitutes voltages of all the buses) and $\boldsymbol{V} = \boldsymbol{Z}_A \boldsymbol{I}_g$ (from (9.4)), $\boldsymbol{V}_g$ is also a function of $\boldsymbol{Z}_A \boldsymbol{I}_g$. Thus, (9.6) can be written as follows:

$$\dot{\boldsymbol{x}} = \boldsymbol{F}(\boldsymbol{x}, \boldsymbol{Z}_A \boldsymbol{I}_g) + \sum_{i=1}^{M} \boldsymbol{g}_i(\boldsymbol{x})u_i. \tag{9.7}$$

In the above equation, as $\boldsymbol{I}_g$ is a function of machine states $\boldsymbol{x}$ and $\boldsymbol{Z}_A$ represents network parameters for a given network configuration, both $\boldsymbol{I}_g$ and $\boldsymbol{Z}_A$ can be consolidated with $\boldsymbol{F}$ to form a new function $\boldsymbol{f}$ as follows:

$$\dot{\boldsymbol{x}} = \boldsymbol{f}(\boldsymbol{x}) + \sum_{i=1}^{M} \boldsymbol{g}_i(\boldsymbol{x})u_i. \tag{9.8}$$

The above equation is the same as (7.78), and hence this establishes the equivalence of power system DAEs in (9.1) and (9.3) and the nonlinear affine ODEs in (7.78) for a given well-defined network configuration. This idea that DAEs and ODEs are equivalent for a given network configuration has also been used to apply nonlinear control theory to power systems in [47–52]. It should be understood that $\boldsymbol{f}$ changes as soon as there is any change in any line parameter or in network structure, e.g., a change in tap position of a transformer, or if there is any change in load at any bus. Even though $\boldsymbol{f}$ may change, the above equivalence remains valid for the new $\boldsymbol{f}$ as long as $\boldsymbol{f}$ remains well defined, that is, the network configuration remains well defined.

To summarize the above equivalence, with V_{ss}^i as input and $\Delta\delta^i$ as output, the power system model given by (9.1)–(9.3) can be represented as (7.78), with various terms defined as follows:

$$
\begin{aligned}
&\boldsymbol{x} = [\boldsymbol{x}^{1^T}\ \boldsymbol{x}^{2^T} \ldots \boldsymbol{x}^{M^T}]^T,\\
&\boldsymbol{x}^i = [\Delta\delta^i\ \Delta\omega^i\ E_d'^i\ E_q'^i\ \Psi_{1d}^i\ \Psi_{2q}^i\ E_{dc}'^i\ V_r^i]^T;\\
&\boldsymbol{f}(\boldsymbol{x}) \text{ is obtained from (9.1)–(9.3) as described above, so we have}\\
&u_i = V_{ss}^i,\\
&\boldsymbol{g}_i(\boldsymbol{x}) = [\boldsymbol{g}_i^{1^T} \boldsymbol{g}_i^{2^T} \ldots \boldsymbol{g}_i^{M^T}],^T\ \boldsymbol{g}_i^i = [0\ 0\ 0\ \frac{K_a^i}{T_{d0}'^i}\ 0\ 0\ 0\ 0]^T,\\
&\boldsymbol{g}_i^j = [0\ 0\ 0\ 0\ 0\ 0\ 0\ 0]^T\ \forall j \neq i,\\
&y_i = h_i(\boldsymbol{x}) = \Delta\delta^i.
\end{aligned}
\tag{9.9}
$$

The normal form for the above multiinput–multioutput (MIMO) representation of power system dynamics is derived as follows.

9.1.1 Relative degree

The relative degree of output $y_i = \Delta\delta^i$ for the MIMO system given by (7.78) and (9.9) is $r_i = 3$, as $r_i = 3$ is the smallest integer for which $L_{\boldsymbol{g}_j} L_{\boldsymbol{f}}^{(r_i-1)} \Delta\delta^i \neq 0$ for at least some $j \in \{1, 2, \ldots, M\}$. That is,

a) $L_{\boldsymbol{g}_j} L_{\boldsymbol{f}}^{(1-1)} \Delta\delta^i = L_{\boldsymbol{g}_j} \Delta\delta^i = 0\ \forall j \in \{1, 2, \ldots, M\}$ and

b) $L_{\boldsymbol{g}_j} L_{\boldsymbol{f}}^{(2-1)} \Delta\delta^i = L_{\boldsymbol{g}_j} L_{\boldsymbol{f}} \Delta\delta^i = L_{\boldsymbol{g}_j} \Delta\omega^i = \frac{K_a^j}{T_{d0}'^j} \frac{\partial \Delta\omega^i}{\partial E_q'^j} = 0\ \forall j \in \{1, 2, \ldots, M\}$, but

c) $L_{g_j}L_f^{(r_i-1)}\Delta\delta^i = L_{g_j}L_f^2\Delta\delta^i = L_{g_j}L_fL_f\Delta\delta^i = L_{g_j}L_f\Delta\omega^i = L_{g_j}[\frac{\omega_b}{2H^i}(T_m^i - T_e^i) - \frac{D^i}{2H^i}\Delta\omega^i] = \frac{-\omega_b K_a^j}{2H^i T_{d0}^{\prime j}}\frac{\partial T_e^i}{\partial E_q^{\prime j}} \neq 0 \ \forall j \in \{1, 2, \ldots, M\}$.

The (i, j) element of the $(M \times M)$ characteristic matrix $\boldsymbol{C}(\boldsymbol{x})$ for the system is $C^{ij}(\boldsymbol{x}) = L_{g_j}L_f^{(r_i-1)}\Delta\delta^i = \frac{-\omega_b K_a^j}{2H^i T_{d0}^{\prime j}}\frac{\partial T_e^i}{\partial E_q^{\prime j}}$. The system has a well-defined relative degree provided that $\boldsymbol{C}(\boldsymbol{x})$ is nonsingular. But $\boldsymbol{C}(\boldsymbol{x})$ is highly complex and depends not only on the states and parameters of all the machines, but also on various loads and line parameters. Thus, it is very difficult to verify the nonsingularity of this matrix. But the existence of a nonsingular $\boldsymbol{C}(\boldsymbol{x})$ is not necessary for the existence of a normal form. In the case of power systems, the existence of a relative degree for each individual output is sufficient to derive the normal form (as described below).

9.1.2 Linearized dynamics

As $r_i = 3$ is the relative degree of $y_i = h_i(\boldsymbol{x}) = \Delta\delta^i$, $r = r_1 + r_2 + \cdots + r_M = 3M \leq n = 8M$. According to Result 7.2(a), $3M$ new states for the linearized dynamics can be defined, with three states for each of the M machines. Using (7.80), the three new states for the ith machine are defined as follows:

$$
\begin{aligned}
z_1^i &= \Delta\delta^i, \ \dot{z}_1^i = L_f\Delta\delta^i = \Delta\dot{\delta}^i = \Delta\omega^i = z_2^i, \\
\dot{z}_2^i &= L_f^2\Delta\delta^i = L_f\Delta\omega^i = \Delta\dot{\omega}^i = z_3^i, \\
\dot{z}_3^i &= L_f^3\Delta\delta^i + \sum_{j=1}^{M} L_{g_j}L_f^2\Delta\delta^i u_j = \Delta\ddot{\omega}^i = v_i.
\end{aligned}
\tag{9.10}
$$

9.1.3 Internal dynamics

The total number of states of the power system is $n = 8M$, and only $3M$ new states are defined in the linearized dynamics. Thus, another $5m$ states need to be defined for the system in order to completely represent the system's dynamics, as explained in Result 7.2(b). One straightforward way to do this is to redefine the states $[E_d^{\prime i} \ \Psi_{2q}^i \ \Psi_{1d}^i \ E_{dc}^{\prime i} \ V_r^i]^T$ of the ith machine as its new states $[w_1^i \ w_2^i \ w_3^i \ w_4^i \ w_5^i]^T = \boldsymbol{w}^i$. It should be noted that the state vector $\boldsymbol{w} = [w_{r+1} \ w_{r+2} \ldots w_n]^T$ in Result 7.2(b) is the same as $\boldsymbol{w} = [\boldsymbol{w}^{1T} \ \boldsymbol{w}^{2T} \ldots \boldsymbol{w}^{MT}]^T$ in this redefinition. This definition is not only simple, but it also has the added advantage that the input $u_i = V_{ss}^i$ does not affect the dynamic equations of these states directly, but only indirectly through the linearized dynamics. In other words, the term $L_{g_j}\phi_i(\boldsymbol{x})$

in (7.81) is zero (as $\frac{\partial E'^i_d}{\partial E'^j_q} = \frac{\partial \Psi^i_{2q}}{\partial E'^j_q} = \frac{\partial \Psi^i_{1d}}{\partial E'^j_q} = \frac{\partial E'^i_{dc}}{\partial E'^j_q} = \frac{\partial V^i_r}{\partial E'^j_q} = 0 \ \forall i, j \in \{1, 2, \ldots, M\}$) and hence, the term $\boldsymbol{p}(\boldsymbol{z}, \boldsymbol{w})\boldsymbol{u}$ in (7.81) vanishes and (7.81) reduces to the following equation:

$$\begin{aligned} \dot{w}^i_1 &= L_f E'^i_d = \dot{E}'^i_d, \quad \dot{w}^i_2 = L_f \Psi^i_{2q} = \dot{\Psi}^i_{2q}, \\ \dot{w}^i_3 &= L_f \Psi^i_{1d} = \dot{\Psi}^i_{1d}, \quad \dot{w}^i_4 = L_f E'^i_{dc} = \dot{E}'^i_{dc}, \\ \dot{w}^i_5 &= L_f V^i_r = \dot{V}^i_r, \quad \dot{\boldsymbol{w}} = \boldsymbol{q}(\boldsymbol{z}, \boldsymbol{w}). \end{aligned} \tag{9.11}$$

Therefore, the eight new states for the ith machine, including both the linearized dynamics and the internal dynamics, are given by $[z^i_1 \ z^i_2 \ z^i_3 \ w^i_1 \ w^i_2 \ w^i_3 \ w^i_4 \ w^i_5]^T = [\Delta\delta^i \ \Delta\omega^i \ \Delta\dot{\omega}^i \ E'^i_d \ \Psi^i_{2q} \ \Psi^i_{1d} \ E'^i_{dc} \ V^i_r]^T$. Also, Eq. (9.11) represents the internal dynamics only when the derivatives $\dot{E}'^i_d$, $\dot{\Psi}^i_{2q}$, $\dot{\Psi}^i_{1d}$, $\dot{E}'^i_{dc}$, and $\dot{V}^i_r$ are represented in terms of the new states. Substituting the expression for I^i_d from (9.2) in the expressions for $\dot{E}'^i_d$ and $\dot{\Psi}^i_{2q}$ from (9.1), followed by substituting the values of K^i_{q1} and K^i_{q2} (using the "List of symbols"), and collecting the coefficients of E'^i_d, Ψ^i_{1d}, and V^i_d, the following equations are obtained:

$$\begin{aligned} \dot{E}'^i_d &= a^i_{11} E'^i_d + a^i_{12} \Psi^i_{2q} + \frac{(X^i_q - X'^i_q)(X''^i_q - X^i_l)}{T'^i_{q0} X''^i_q (X'^i_q - X^i_l)} V^i_d, \\ \dot{\Psi}^i_{2q} &= a^i_{21} E'^i_d + a^i_{22} \Psi^i_{2q} + \frac{-(X'^i_q - X^i_l)}{T''^i_{q0} X''^i_q} V^i_d, \quad \text{where} \\ a^i_{11} &= \frac{-1}{T'^i_{q0}} \left[1 + \frac{(X^i_q - X'^i_q)(X^{i2}_l + X'^i_q X''^i_q - 2X''^i_q X^i_l)}{X''^i_q (X'^i_q - X^i_l)^2} \right], \\ a^i_{12} &= \frac{-X^i_l (X^i_q - X'^i_q)(X'^i_q - X''^i_q)}{T'^i_{q0} X''^i_q (X'^i_q - X^i_l)^2}, \\ a^i_{21} &= \frac{-X^i_l}{T''^i_{q0} X''^i_q}, \quad a^i_{22} = \frac{-X'^i_q}{T''^i_{q0} X''^i_q}. \end{aligned} \tag{9.12}$$

Similarly, substituting the expressions for I^i_q, K^i_{d1}, and K^i_{d2} in the expression for $\dot{\Psi}^i_{1d}$ from (9.1) and collecting the coefficients of E'^i_d, Ψ^i_{1d}, and V^i_q, we find the following equation:

$$\dot{\Psi}^i_{1d} = \frac{X^i_l}{T''^i_{d0} X''^i_d} E'^i_q + \frac{-X'^i_d}{T''^i_{d0} X''^i_d} \Psi^i_{1d} + \frac{X'^i_d - X^i_l}{T''^i_{d0} X''^i_d} V^i_q. \tag{9.13}$$

As $E_q^{\prime i}$ is not equal to any of the new states, it should be expressed in terms of the new states in the above equation. Substituting the expressions for I_d^i and I_q^i from (9.2) in the expression for T_e^i from (9.1), T_e^i becomes as follows:

$$T_e^i = \frac{V_q^i}{\omega^i}\frac{E_d^{\prime i}K_{q1}^i - \Psi_{2q}^i K_{q2}^i - V_d^i}{X_q^{\prime\prime i}/\omega_b} + \frac{V_d^i}{\omega^i}\frac{V_q^i - E_q^{\prime i}K_{d1}^i - \Psi_{1d}^i K_{d2}^i}{X_d^{\prime\prime i}/\omega_b}. \tag{9.14}$$

Substituting the above expression for T_e^i in the expression for $\Delta\dot{\omega}^i$ from (9.1), the following equations are derived:

$$\begin{aligned}
\Delta\dot{\omega}^i =& \frac{\omega_b}{2H^i}T_m^i - \frac{\omega_b{}^2 V_q^i}{2H^i\omega^i X_q^{\prime\prime i}}(E_d^{\prime i}K_{q1}^i - \Psi_{2q}^i K_{q2}^i - V_d^i) \\
& - \frac{\omega_b{}^2 V_d^i}{2H^i\omega^i X_d^{\prime\prime i}}(V_q^i - E_q^{\prime i}K_{d1}^i - \Psi_{1d}^i K_{d2}^i) - \frac{D^i}{2H^i}\Delta\omega^i \\
\Rightarrow E_q^{\prime i} =& \frac{X_d^{\prime\prime i}V_q^i K_{q1}^i}{X_q^{\prime\prime i}V_d^i K_{d1}^i}[E_d^{\prime i} - \frac{K_{q2}^i}{K_{q1}^i}\Psi_{2q}^i] - \frac{K_{d2}^i}{K_{d1}^i}\Psi_{1d}^i + \frac{X_q^{\prime\prime i} - X_d^{\prime\prime i}}{X_q^{\prime\prime i}}V_q^i \\
& + \frac{2H^i\omega^i X_d^{\prime\prime i}}{\omega_b{}^2 V_d^i K_{d1}^i}[\Delta\dot{\omega}^i + \frac{D^i}{2H^i}\Delta\omega^i - \frac{\omega_b}{2H^i}T_m^i].
\end{aligned}$$

Plugging the above expression for $E_q^{\prime i}$ in (9.13) and collecting the coefficients of Ψ_{1d}^i and V_q^i, we obtain the following equation:

$$\begin{aligned}
\dot{\Psi}_{1d}^i =& a_{31}^i E_d^{\prime i} + a_{32}^i\Psi_{2q}^i + a_{33}^i\Psi_{1d}^i + [\frac{X_d^{\prime i}}{X_d^{\prime\prime i}} - \frac{X_l^i}{X_q^{\prime\prime i}}]\frac{V_q^i}{T_{d0}^{\prime\prime i}} \\
& + \frac{2H^i\omega^i X_l^i(X_d^\prime - X_l)}{T_{d0}^{\prime\prime i}\omega_b{}^2 V_d^i(X_d^{\prime\prime} - X_l)}[\Delta\dot{\omega}^i + \frac{D^i}{2H^i}\Delta\omega^i - \frac{\omega_b}{2H^i}T_m^i], \\
\text{where } a_{31}^i =& \frac{-X_l^{\prime i}\tan(\delta_0^i + \Delta\delta^i - \theta^i)(X_q^{\prime\prime i} - X_l^i)(X_d^{\prime i} - X_l^i)}{T_{d0}^{\prime\prime i}X_q^{\prime\prime i}(X_q^{\prime i} - X_l^i)(X_d^{\prime\prime i} - X_l^i)},
\end{aligned} \tag{9.15}$$

$$a_{32}^i = \frac{-(X_q^{\prime i} - X_q^{\prime\prime i})}{(X_q^{\prime\prime i} - X_l^i)}a_{31}^i, \quad a_{33}^i = \frac{-(X_d^{\prime i} - X_l^i)}{T_{d0}^{\prime\prime i}}.$$

Next, the expression for I_q^i from (9.2) is substituted in the expression for $\dot{E}_{dc}^{\prime i}$ from (9.1) to obtain the following equation:

$$\dot{E}_{dc}^{\prime i} = a_{41}^i E_d^{\prime i} + a_{42}^i \Psi_{2q}^i - \frac{1}{T_c^i} E_{dc}^{\prime i} - \frac{(X_d^{\prime\prime i} - X_q^{\prime\prime i})}{T_c^i X_q^{\prime\prime i}} V_d^i,$$
$$\text{where } a_{41}^i = \frac{K_{q1}^i (X_d^{\prime\prime i} - X_q^{\prime\prime i})}{T_c^i X_q^{\prime\prime i}}, \quad a_{42}^i = \frac{-K_{q2}^i (X_d^{\prime\prime i} - X_q^{\prime\prime i})}{T_c^i X_q^{\prime\prime i}}. \tag{9.16}$$

Finally, using the expressions for $\dot{E}_d^{\prime i}$, $\dot{\Psi}_{2q}^i$, $\dot{\Psi}_{1d}^i$, $\dot{E}_{dc}^{\prime i}$, and $\dot{V}_r^i$ (from (9.12), (9.15), (9.16), and (9.1)) in (9.11) and replacing $\Delta\delta^i$, $\Delta\omega^i$, $\Delta\dot{\omega}^i$, $E_d^{\prime i}$, Ψ_{2q}^i, Ψ_{1d}^i, $E_{dc}^{\prime i}$, and V_r^i with z_1^i, z_2^i, z_3^i, w_1^i, w_2^i, w_3^i, w_4^i, and w_5^i, respectively, the internal dynamics are as follows:

$$\dot{\boldsymbol{w}}^i = \boldsymbol{a}^i \boldsymbol{w}^i + \boldsymbol{b}^i, \ \forall\, i \in \{1, 2, \ldots, M\}, \text{ where}$$

$$\boldsymbol{a}^i = \begin{bmatrix} a_{11}^i & a_{12}^i & 0 & 0 & 0 \\ a_{21}^i & a_{22}^i & 0 & 0 & 0 \\ a_{31}^i & a_{32}^i & a_{33}^i & 0 & 0 \\ a_{41}^i & a_{42}^i & 0 & \frac{-1}{T_c^i} & 0 \\ 0 & 0 & 0 & 0 & \frac{-1}{T_r^i} \end{bmatrix}, \quad \boldsymbol{b}^i = \begin{bmatrix} b_1^i \\ b_2^i \\ b_3^i \\ b_4^i \\ b_5^i \end{bmatrix},$$

$$\boldsymbol{w}^i = [w_1^i \ w_2^i \ w_3^i \ w_4^i \ w_5^i]^T,$$

$$b_1^i = \frac{-(X_q^i - X_q^{\prime i})(X_q^{\prime\prime i} - X_l^i)}{T_{q0}^{\prime i} X_q^{\prime\prime i}(X_q^{\prime i} - X_l^i)} V^i \sin(\delta_0^i + z_1^i - \theta^i), \tag{9.17}$$

$$b_2^i = \frac{(X_q^{\prime i} - X_l^i)}{T_{q0}^{\prime\prime i} X_q^{\prime\prime i}} V^i \sin(\delta_0^i + z_1^i - \theta^i),$$

$$b_3^i = \frac{-2H^i(\omega_b + z_2^i) X_l^i (X_d^{\prime} - X_l)}{T_{d0}^{\prime\prime i} \omega_b{}^2 V^i \sin(\delta_0^i + z_1^i - \theta^i)(X_d^{\prime\prime} - X_l)} [z_3^i + \frac{D^i}{2H^i} z_2^i - \frac{\omega_b}{2H^i} T_m^i] + [\frac{X_d^{\prime i}}{X_d^{\prime\prime i}} - \frac{X_l^i}{X_q^{\prime\prime i}}] \frac{V^i \cos(\delta_0^i + z_1^i - \theta^i)}{T_{d0}^{\prime\prime i}},$$

$$b_4^i = \frac{(X_d^{\prime\prime i} - X_q^{\prime\prime i})}{T_c^i X_q^{\prime\prime i}} V^i \sin(\delta_0^i + z_1^i - \theta^i), \quad b_5^i = \frac{1}{T_r^i} V^i.$$

Elements of $\boldsymbol{a}^i$ are as in (9.12)–(9.16); V^i, θ^i are as in (9.3) or (9.4).

Thus, (9.10) and (9.17) completely specify a normal form for the power system dynamics.

9.2 ASYMPTOTIC STABILITY OF ZERO DYNAMICS

The power system will be asymptotically stable under a nonlinear control method based on the normal form if both its linearized dynamics and its internal dynamics are asymptotically stable, as explained in Section 7.2.1. The asymptotic stability of linearized dynamics can be ensured using any desired control method (provided that the closed-loop linearized dynamics are not already unstable, or the operating condition has not crossed some stability threshold; for example, any system fault should be cleared before its critical clearing time (CCT)). From Result 7.3, if the closed-loop linearized dynamics are stable, then the internal dynamics will be asymptotically stable if and only if the zero dynamics are asymptotically stable.

Theorem 9.1. *The zero dynamics for a power system (with normal form given by* (9.10) *and* (9.17)*) are asymptotically stable irrespective of the operating condition.*

Proof. Using Result 7.3, the zero dynamics of a power system for a given initial operating condition are obtained from its internal dynamics (given by (9.17)) by putting $\boldsymbol{z} = \mathbf{0}$, that is, by putting $z_1^i = 0$, $z_2^i = 0$, and $z_3^i = 0$ for all $1 \leq i \leq M$. After substituting $\Delta\dot{\omega}^i$ and $\Delta\omega^i$ with z_3^i and z_2^i, respectively, the expression for the zero dynamics for $\Delta\dot{\omega}^i$ from (9.1) reduces to $z_3^i = \frac{\omega_b}{2H^i}[T_m^i - T_e^i] - \frac{D^i}{2H^i}z_2^i \Rightarrow T_m^i - T_e^i = 0$ (as $z_2^i = 0$ and $z_3^i = 0$). Thus, in zero dynamics, T_m^i and T_e^i are equal. They can be equal (for all $1 \leq i \leq M$) only if the system loads and generations are exactly matched and the line connections and parameters are constant. Thus, $\boldsymbol{P}$, $\boldsymbol{Q}$, $\boldsymbol{B}$, and $\boldsymbol{G}$ in (9.3) remain constant in zero dynamics, and hence V^i and θ^i also remain constant for all $1 \leq i \leq N$. The elements of $\boldsymbol{b}^i$ in (9.17) also remain constant for zero dynamics, as they are functions of z_1^i, z_2^i, z_3^i, V^i, and θ^i. As $\boldsymbol{b}^i$ is a constant for all $1 \leq i \leq M$, the zero dynamics will be asymptotically stable if and only if all the eigenvalues of $\boldsymbol{a}^i$ are negative for all $1 \leq i \leq M$ [77]. The eigenvalues of $\boldsymbol{a}^i$ are the roots of its following characteristic polynomial:

$$\begin{aligned}&\text{determinant}(\boldsymbol{a}^i - \lambda\boldsymbol{I}) = 0\\&\Rightarrow [[a_{11}^i - \lambda][a_{22}^i - \lambda] - a_{12}^i a_{21}^i][a_{33}^i - \lambda][\frac{-1}{T_c^i} - \lambda][\frac{-1}{T_r^i} - \lambda] = 0.\end{aligned} \tag{9.18}$$

Three roots of the above equation are $\lambda = a_{33}^i$, $\lambda = \frac{-1}{T_c^i}$, $\lambda = \frac{-1}{T_r^i}$, and the other two roots are the solutions of the following equation:

$$\begin{aligned}&\lambda^2 + b\lambda + c = 0,\ b = -a_{11}^i - a_{22}^i,\ c = a_{11}^i a_{22}^i - a_{12}^i a_{21}^i\\&\Rightarrow \lambda = (-b \pm \sqrt{b^2 - 4c})/2 = -(b \mp \sqrt{b^2 - 4c})/2.\end{aligned} \tag{9.19}$$

Using (9.12), c in the above equation is evaluated as follows:

$$
\begin{aligned}
c = & \frac{-1}{T'^{i}_{q0}}\left[1+\frac{(X^{i}_{q}-X'^{i}_{q})(X^{i2}_{l}+X'^{i}_{q}X''^{i}_{q}-2X''^{i}_{q}X^{i}_{l})}{X''^{i}_{q}(X'^{i}_{q}-X^{i}_{l})^{2}}\right]\frac{-X'^{i}_{q}}{T''^{i}_{q0}X''^{i}_{q}} \\
& -\frac{-X^{i}_{l}(X^{i}_{q}-X'^{i}_{q})(X'^{i}_{q}-X''^{i}_{q})}{T'^{i}_{q0}X''^{i}_{q}(X'^{i}_{q}-X^{i}_{l})^{2}}\frac{-X^{i}_{l}}{T''^{i}_{q0}X''^{i}_{q}} = \frac{X'^{i}_{q}}{T'^{i}_{q0}T''^{i}_{q0}X''^{i}_{q}} \\
& +\frac{(X^{i}_{q}-X'^{i}_{q})(X'^{i}_{q}(X^{i2}_{l}-2X''^{i}_{q}X^{i}_{l}+X'^{i}_{q}X''^{i}_{q})-X^{i2}_{l}(X'^{i}_{q}-X''^{i}_{q}))}{T'^{i}_{q0}T''^{i}_{q0}X''^{i2}_{q}(X'^{i}_{q}-X^{i}_{l})^{2}} \\
= & \frac{X'^{i}_{q}}{T'^{i}_{q0}T''^{i}_{q0}X''^{i}_{q}}+\frac{(X^{i}_{q}-X'^{i}_{q})(X'^{i2}_{q}X''^{i}_{q}+X^{i2}_{l}X''^{i}_{q}-2X'^{i}_{q}X^{i}_{l}X''^{i}_{q})}{T'^{i}_{q0}T''^{i}_{q0}X''^{i2}_{q}(X'^{i}_{q}-X^{i}_{l})^{2}} \\
= & \frac{X'^{i}_{q}}{T'^{i}_{q0}T''^{i}_{q0}X''^{i}_{q}}+\frac{(X^{i}_{q}-X'^{i}_{q})(X'^{i}_{q}-X^{i}_{l})^{2}X''^{i}_{q}}{T'^{i}_{q0}T''^{i}_{q0}X''^{i2}_{q}(X'^{i}_{q}-X^{i}_{l})^{2}} \\
= & \frac{X'^{i}_{q}}{T'^{i}_{q0}T''^{i}_{q0}X''^{i}_{q}}+\frac{X^{i}_{q}-X'^{i}_{q}}{T'^{i}_{q0}T''^{i}_{q0}X''^{i}_{q}} = \frac{X^{i}_{q}}{T'^{i}_{q0}T''^{i}_{q0}X''^{i}_{q}} > 0.
\end{aligned}
$$

Also, as explained in [2], the following inequalities hold:

$$X^{i}_{q} > X'^{i}_{q} \geq X'^{i}_{d} > X''^{i}_{q} > X^{i}_{l} > 0. \tag{9.20}$$

Hence, in (9.12), $a^{i}_{11} < 0$, $a^{i}_{12} < 0$, $a^{i}_{21} < 0$, and $a^{i}_{22} < 0$, and hence, $b = -a^{i}_{11} - a^{i}_{22} > 0$ and $a^{i}_{12}a^{i}_{21} > 0$. This also implies that $b^2 - 4c = (a^{i}_{11} + a^{i}_{22})^2 - 4(a^{i}_{11}a^{i}_{22} - a^{i}_{12}a^{i}_{21}) = (a^{i}_{11} - a^{i}_{22})^2 + 4(a^{i}_{12}a^{i}_{21}) > 0$, and thus $\sqrt{b^2 - 4c}$ is a real number and, hence, is positive, and the following inequalities exist:

$$
\begin{aligned}
& c > 0,\ b > 0,\ \sqrt{b^2-4c} > 0 \\
& \Rightarrow b^2 > b^2 - 4c \Rightarrow b > \sqrt{b^2-4c} > 0 \\
& \Rightarrow (b \mp \sqrt{b^2-4c})/2 > 0 \Rightarrow -(b \mp \sqrt{b^2-4c})/2 < 0.
\end{aligned}
\tag{9.21}
$$

From (9.19) and (9.21), $\lambda = -(b \mp \sqrt{b^2-4c})/2 < 0$ are two roots of (9.18) and the other roots are $\lambda = a^{i}_{33} = \frac{-(X'^{i}_{d}-X^{i}_{l})}{T''^{i}_{d0}} < 0$ (using (9.15) and (9.20)), $\lambda = \frac{-1}{T^{i}_{c}} < 0$, and $\lambda = \frac{-1}{T^{i}_{r}} < 0$. Thus all the five roots of (9.18) are negative, and $\boldsymbol{a}^{i}$ has negative eigenvalues for all $1 \leq i \leq M$. Therefore the zero dynamics are asymptotically stable. □

It should be noted that the above Lemma does not consider, and is not applicable to, the case when the power system state is at the origin, that is, when the power system is completely "off", with $T_m = T_e = 0$ for all the machines.

9.3 OVERALL STABILITY AND CONTROL EXPRESSION

The internal dynamics of a power system are asymptotically stable (from Lemma 9.1 and Result 7.3), and hence the overall dynamics of the system will be asymptotically stable if its linearized dynamics are stabilized using adequate control. The linearized dynamics from (9.10) can be written as follows:

$$\begin{bmatrix} \dot{z}_1^i \\ \dot{z}_2^i \\ \dot{z}_3^i \end{bmatrix} = \begin{bmatrix} 0 & 1 & 0 \\ 0 & 0 & 1 \\ 0 & 0 & 0 \end{bmatrix} \begin{bmatrix} z_1^i \\ z_2^i \\ z_3^i \end{bmatrix} + \begin{bmatrix} 0 \\ 0 \\ 1 \end{bmatrix} v_i, \quad 1 \leq i \leq M. \tag{9.22}$$

The above linear dynamics can be asymptotically stabilized in an optimal manner using the standard LQR, in which a linear feedback of states is found and given as input (that is, $v_i = [K_1^i \; K_2^i \; K_3^i][z_1^i \; z_2^i \; z_3^i]^T = \boldsymbol{K}^i \boldsymbol{z}^i$) such that the sum of the weighted quadratic costs corresponding to state(s) and input(s) (given by $\sum_{k=1}^{\infty}\{\boldsymbol{z}^i(k)^T \mathbf{Q}_i \boldsymbol{z}^i(k) + v_i(k)^2 R_i\}$) is minimized [77]. The state weighting matrix $\mathbf{Q}_i$ and input weight R_i are found using trial-and-error to fine-tune the control performance, and in this chapter these are taken as $\mathbf{Q}_i = \text{diag}(50, 100, 10)$ and $R_i = 1$, for all $1 \leq i \leq M$. With these weights and with the state matrix and input matrix as given by (9.22), the LQR feedback law is found to be as follows:

$$\begin{aligned} v_i &= -7.0711 z_1^i - 13.652 z_2^i - 6.1077 z_3^i \\ \Rightarrow \Delta\ddot{\omega}^i &= -(7.0711\Delta\delta^i + 13.652\Delta\omega^i + 6.1077\Delta\dot{\omega}^i). \end{aligned} \tag{9.23}$$

Also, from (9.1), $\Delta\ddot{\omega}^i = -\dfrac{\omega_b}{2H^i}\dot{T}_e^i - \dfrac{D^i}{2H^i}\Delta\dot{\omega}^i$.

Differentiating both sides of (9.14) and using the identities $E_q'^i K_{d1}^i + \Psi_{1d}^i K_{d2}^i = V_q^i - I_d^i X_d''^i$ and $E_d'^i K_{q1}^i - \Psi_{2q}^i K_{q2}^i = V_d^i + I_q^i X_q''^i$ (from (9.2)), $\dot{T}_e^i$ is obtained as follows:

$$\dot{T}_e^i = T_e^{i'} - \dot{E}_q^{\prime i}\frac{\omega_b V_d^i K_{d1}^i}{\omega^i X_d^{\prime\prime i}}, \text{ where}$$

$$\begin{aligned} T_e^{i'} =& \dot{E}_d^{\prime i}\frac{\omega_b V_q^i K_{q1}^i}{\omega^i X_q^{\prime\prime i}} - \dot{\Psi}_{2q}^i\frac{\omega_b V_q^i K_{q2}^i}{\omega^i X_q^{\prime\prime i}} - \dot{\Psi}_{1d}^i\frac{\omega_b V_d^i K_{d2}^i}{\omega^i X_d^{\prime\prime i}} \\ &+ \dot{V}_q^i\frac{\omega_b}{\omega^i}[I_q^i + \frac{V_d^i}{X_d^{\prime\prime i}}] + \dot{V}_d^i\frac{\omega_b}{\omega^i}[I_d^i - \frac{V_q^i}{X_q^{\prime\prime i}}] - \Delta\dot{\omega}^i\frac{T_e^i}{\omega^i}. \end{aligned} \tag{9.24}$$

Also, $\dot{E}_q^{\prime i}$ from (9.2) can be written as follows:

$$\begin{aligned} \dot{E}_q^{\prime i} =& \frac{1}{T_{d0}^{\prime i}}[E_{fd}^i + E_q^{i\prime\prime}], \text{ where} \\ E_q^{i\prime\prime} =& (X_d^i - X_d^{\prime i})[K_{d1}^i I_d^i + K_{d2}^i\frac{\Psi_{1d}^i - E_q^{\prime i}}{X_d^{\prime i} - X_l^i}] - E_q^{\prime i}. \end{aligned} \tag{9.25}$$

Finally, using (9.1), (9.23)–(9.25), the optimal V_{ss}^i becomes as follows:

$$\begin{aligned} V_{ss}^i =& \frac{E_{fd}^i}{K_a^i} - V_{ref}^i + V_r^i, \ E_{fdmin}^i \leq E_{fd}^i \leq E_{fdmax}^i, \\ E_{fd}^i =& \frac{T_e^{i'} - [7.07\Delta\delta^i + 13.65\Delta\omega^i + [6.11 - \frac{D^i}{2H^i}]\Delta\dot{\omega}^i]\frac{2H^i}{\omega_b}}{(\omega_b V_d^i K_{d1}^i)/(\omega^i X_d^{\prime\prime i} T_{d0}^{\prime i})} - E_q^{i\prime\prime}, \end{aligned} \tag{9.26}$$

where $T_e^{i'}$, $E_q^{i''}$ are defined in (9.24), (9.25), respectively.

Remark 9.1. Although the normal form for power system dynamics, their asymptotic stability, and the final control expression have been derived using a static excitation system, these can be similarly derived considering other types of excitation systems as well.

Remark 9.2. The control input V_{ss}^i in (9.26) is a small-signal input given to a machine's excitation system. It should be zero in steady state and should be nonzero only for around 10–20 s after a disturbance (which is the time frame of interest for rotor angle stability). Thus, V_{ss}^i should be passed through a washout filter to filter out any direct-current (DC) component in it and then be given as an input to the excitation system. The largest time constant that describes the dynamics corresponding to rotor angle stability is given by T'_{d0}, which is around 3–10 s for a thermal unit in a power system [2]. For the machines considered in Section 9.5, T'_{d0} is around 5 s. Also, the time period of any oscillatory mode which is of interest (including interarea modes) is not more than 5 s. Thus, the time constant of the washout filter is taken as 5 s, so that any slow dynamics which have time constants higher than 5 s (or any DC components) are filtered out.

As the washout filter removes any DC components in the signal and as $\Delta\dot{\omega}^i = (\dot{\omega}^i - \dot{\omega}^i_b) = \dot{\omega}^i$, $\Delta\omega^i = (\omega^i - \omega^i_b)$, and $\Delta\delta^i = (\delta^i - \delta^i_0)$ are DC signals in the postfault equilibrium, when the fault is removed and the system stabilizes to a new steady state (and hence to a new topology), these DC components are also filtered out by the washout filter. Thus postfault rotor angle and rotor velocity can be different from their prefault values, and the postfault dynamics can easily adjust to any topological changes in the system. Also, due to filtering out of the DC components by the washout filter, any initial value of δ^i_0 can be taken, and in this chapter it is taken as 0 rad.

Remark 9.3. The stability criteria given in Section 9.2 and Section 9.3 will remain valid even if saturation of the unit is considered. This is because saturation in synchronous machines is manifested as changes in its parameters (see [126]), that is, changes in X^i_d, X^i_q, X'^i_d, X'^i_q, and so on. Thus, even if these parameters get altered because of saturation, the relation given by (9.20) still remains valid, and hence the stability criteria given in Section 9.2 and Section 9.3 are valid.

9.4 DECENTRALIZED DYNAMIC STATE ESTIMATION

Using the decentralized DSE method described in previous chapters, the estimates of $\Delta\delta^i$, $\Delta\omega^i$, E'^i_q, E'^i_d, Ψ^i_{2q}, Ψ^i_{1d}, V^i_r are found for the ith machine using just the voltage and current phasors measured at that machine. These estimates are then used in (9.26) to find the input V^i_{ss} for the machine's excitation system at each time sample. The estimate of the derivative of a term used in (9.26) is found by subtracting the estimate of the term in the previous sample from its estimate in the current sample and dividing the difference by the sampling period. For example, the estimate of $\dot{V}^i_q(k)$ is $\frac{V^i_q(k)-V^i_q(k-1)}{T_0}$, where k denotes the kth time sample, T_0 is the sampling period, and $V^i_q(k) = V^i(k)\cos(\delta^i_0 + \Delta\delta^i(k) - \theta^i(k))$ (from (9.1)).

Remark 9.4. As V^i_d and V^i_q are algebraic quantities, their time derivatives can have very large magnitudes during switching events or any other disturbances, and hence these large magnitude derivatives should be detected and filtered out as they can jeopardize the stability of the system. In this chapter, this is done by comparing the derivative at the kth time sample with the derivative at the $(k-1)$th time sample, and if the difference between the two is more than a predetermined value, then the derivative at the kth time sample is set to zero. In this chapter, this predetermined value is taken as 0.04, as this value successfully removes all such large magnitude derivatives and at the same time retains other derivatives. This value is found using trial-and-error.

9.5 CASE STUDY

In rest of the chapter, the developed nonlinear control method based on DSE and normal forms is referred to as the DSE-normal-form (DSE-NF) method. The 68-bus test system (Fig. A.1) has been used for the case study and MATLAB-Simulink has been used for its modeling and simulation. A detailed description of the system is available in Appendix A. The following two cases have been considered to assess the performance of the DSE-NF nonlinear control method for both small signal stability and transient stability.

9.5.1 Case A: Assessment of small signal stability

In this case, the test system starts from steady state, and at $t = 1$ s a small disturbance takes place in which one of the tie lines of the double circuit line between buses 53–54 goes out of service. The test system has four interarea modes in the range 0.1–1.0 Hz and three of these modes are very poorly damped with damping ratios less than 3% (Table 9.1). Thus, after the small disturbance, poorly damped oscillations ensue in the open-loop system, which need to be controlled using an adequate method of control. Three different control methods have been considered to control the oscillations: (1) using power system stabilizer (PSS) control, (2) using decentralized linear control, and (3) using the DSE-NF nonlinear control. The employed decentralized linear control is based on the extended linear quadratic regulator (ELQR) control described in Chapter 8. A description of the PSS control method and its parameters is also available in Appendix A.

The plots of the power flow in the interarea line between buses 60–61 for the three control types are shown in Fig. 9.1 and the plots for δ, ω, V, and E_{fd} of units 1 and 8 are shown in Fig. 9.2 and Fig. 9.3. Units 1 and 8 have been chosen for showing the plots because the disturbance takes close

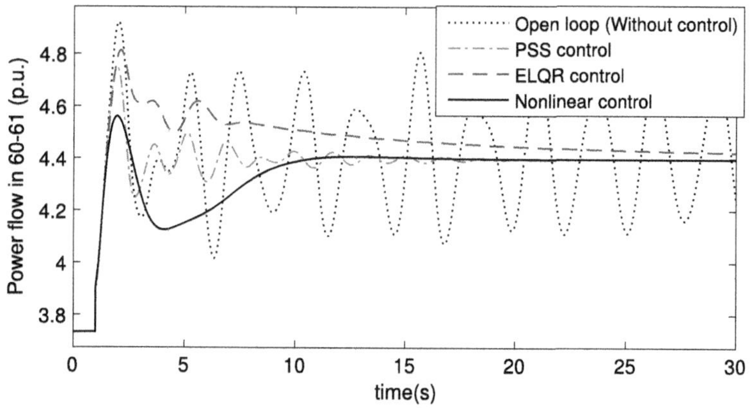

Figure 9.1 Small signal stability comparison: Power flow in interarea line.

to these units. Also, the plots for a unit which is not close to the disturbance (unit 13) are shown in Fig. 9.4. It can be observed in Figs. 9.1–9.4 that the damping provided to the oscillations by the DSE-NF nonlinear control is better than that provided by the other two controls. Also, the voltage regulation provided by the DSE-NF nonlinear control is on a par with the other two methods.

9.5.1.1 Modal and sensitivity analysis

Table 9.1 presents the modal analysis of interarea modes, wherein the frequencies and damping ratios of the three poorly damped interarea modes are shown. It can be seen in Table 9.1 that the damping ratios for the DSE-NF nonlinear control are much higher as compared to the other two control methods, while the corresponding frequencies for the DSE-NF nonlinear control are much lower. To understand the reason behind this, further analysis of the sensitivity of the interarea modes to system states needs to be done.

The sensitivity of a system's *a*th mode (or eigenvalue) to the system's *b*th state is given by the participation factor P_{ba} and is equal to the product of the *b*th entry of the *a*th right eigenvector and the *b*th entry of the *a*th left eigenvector [2]. As the interarea modes are the most significant electromechanical modes of the test system, it is logical to see how sensitive these modes are to the various electromechanical states of the system (electromechanical states are the δ and ω of various generating units in the system). Table 9.2 shows the top three normalized participation factors (NPFs) of

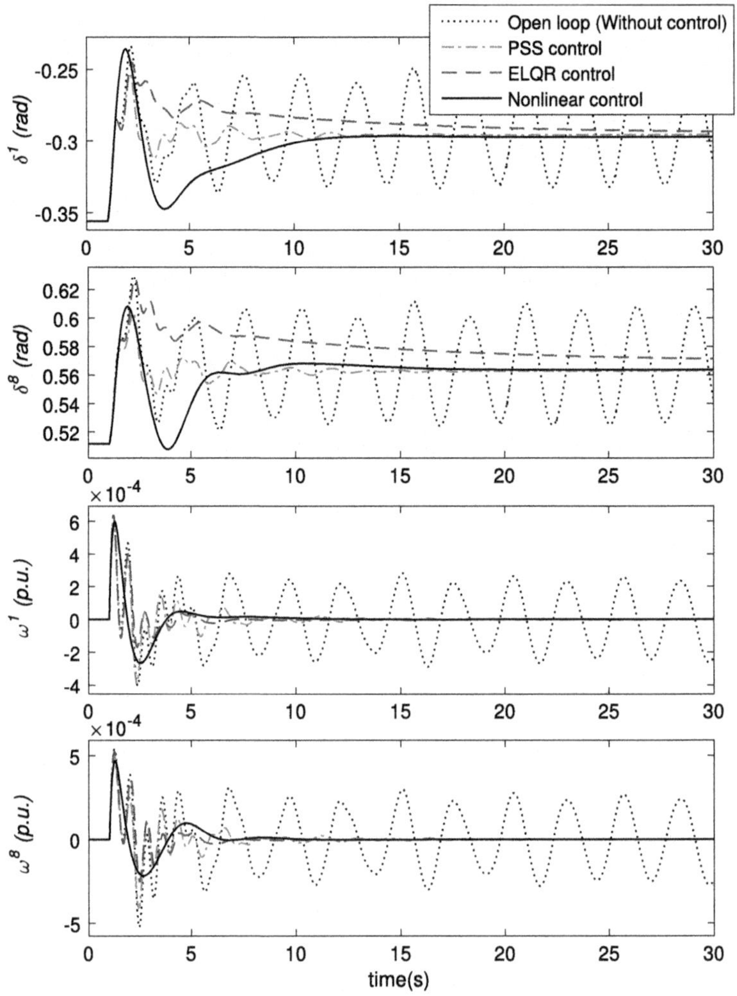

Figure 9.2 Small signal stability comparison: δ and ω for units 1 and 8.

the electromechanical states in each of the interarea modes for the four control scenarios.

It can be seen that the participation in the three interarea modes from various electromechanical states is among the highest for three scenarios: open-loop, PSS control, and ELQR control. But this participation is significantly reduced for the DSE-NF nonlinear control. Instead, the highest participating states in the case of nonlinear control are various controller states, as can be seen in Table 9.3 (the hat, ^, denotes

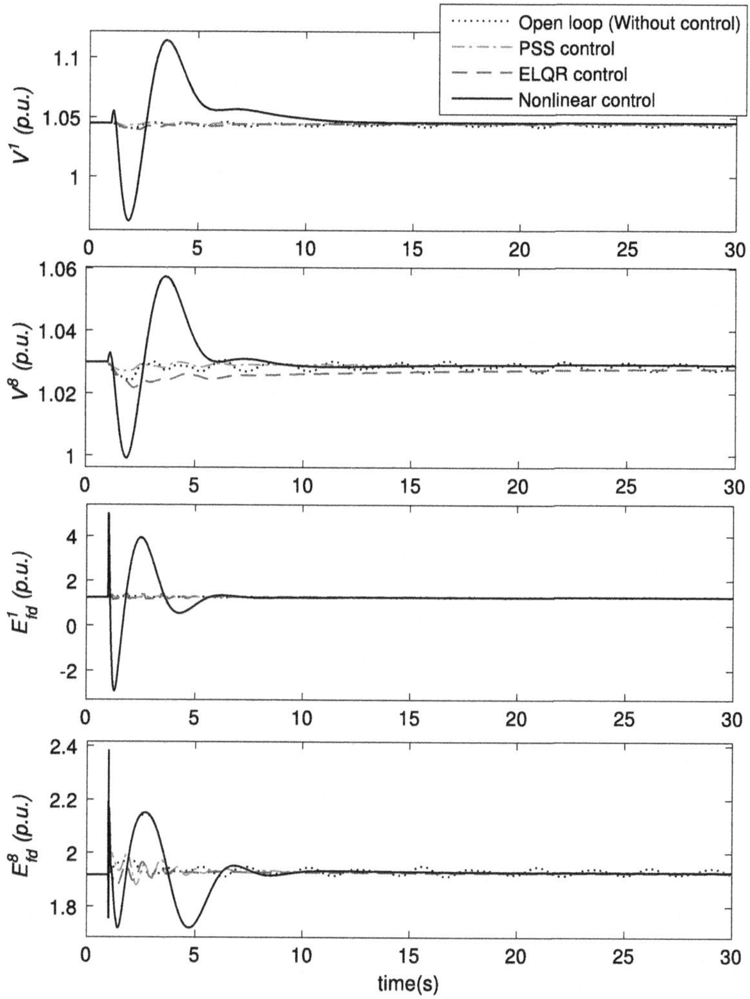

Figure 9.3 Small signal stability comparison: V and E_{fd} for units 1 and 8.

that the quantities are estimated controller states). This high participation of controller states in the interarea modes is the reason behind relatively low frequencies and relatively high damping ratios for the interarea modes.

The overall system damping is decided not only by the frequencies of the dominant modes, but also by the damping ratios of those modes [2]. Thus, the reduction in frequencies of the interarea modes in the case of nonlinear control has no adverse effect on the control performance because

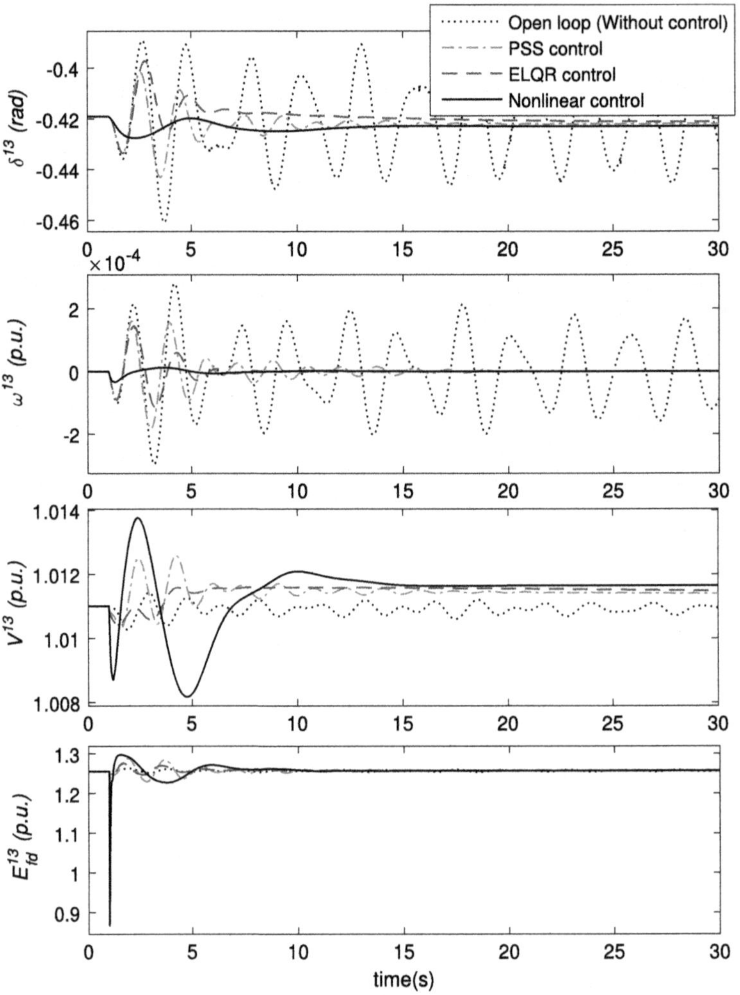

Figure 9.4 Small signal stability comparison for unit 13.

this reduction is well compensated by a significant increase in damping ratios, and the overall damping provided by the DSE-NF nonlinear control is better than the other two control methods.

Remark 9.5. *Performance of other methods of nonlinear control*: Other methods of nonlinear control have also been tested for controlling the oscillations caused by the aforementioned small disturbance in the test system. These methods include control using two methods based on normal forms given

Table 9.1 Modal analysis of interarea modes

	Open-loop	PSS control	ELQR control	Nonlinear control
Mode 1 frequency (Hz)	0.39	0.43	0.31	0.12
Mode 1 damping ratio (%)	0.9	11.4	18.7	47.2
Mode 2 frequency (Hz)	0.52	0.54	0.47	0.20
Mode 2 damping ratio (%)	2.1	6.3	9.8	92.8
Mode 3 frequency (Hz)	0.60	0.63	0.54	0.21
Mode 3 damping ratio (%)	1.2	5.7	11.0	91.0

Table 9.2 Sensitivity analysis: Normalized participation factors (NPFs) of electromechanical states in the three interarea modes

	Mode 1		Mode 2		Mode 3	
	State	NPF	State	NPF	State	NPF
Open-loop	δ^{15}	1.00	ω^{16}	1.00	ω^{13}	1.00
	ω^{15}	0.99	δ^{16}	0.99	δ^{13}	0.99
	δ^{14}	0.84	δ^{14}	0.81	ω^{16}	0.20
PSS control	ω^{13}	1.00	ω^{16}	1.00	ω^{13}	1.00
	ω^{15}	0.89	δ^{16}	0.93	δ^{13}	0.92
	δ^{15}	0.83	ω^{14}	0.73	ω^{16}	0.13
ELQR control	δ^{1}	1.00	δ^{16}	1.00	ω^{13}	1.00
	ω^{15}	0.96	ω^{16}	0.97	δ^{13}	0.98
	δ^{15}	0.94	ω^{14}	0.94	δ^{16}	0.32
Nonlinear control	δ^{14}	0.30	ω^{15}	0.25	ω^{13}	0.39
	δ^{15}	0.27	ω^{14}	0.10	ω^{12}	0.10
	δ^{16}	0.26	ω^{16}	0.07	ω^{1}	0.10

Table 9.3 Top three normalized participation factors (NPFs) of system states in the interarea modes for the DSE-NF nonlinear control

	Mode 1		Mode 2		Mode 3	
	State	NPF	State	NPF	State	NPF
Nonlinear control	$\hat{V}_q^{16}$	1.00	$\hat{V}_q^{15}$	1.00	$\hat{V}_q^{13}$	1.00
	$\hat{V}_d^{16}$	0.75	$\hat{\Psi}_{1d}^{15}$	0.61	$\hat{\omega}^{13}$	0.80
	$\hat{\Psi}_{1d}^{16}$	0.63	$\hat{E}_q'^{15}$	0.52	$\hat{\Psi}_{1d}^{13}$	0.76

in [47] and [48] and control using two Lyapunov-based methods given in [53] and [56]. It has been found that the methods either destabilize the system or produce unwanted oscillations (power flow in line 60–61

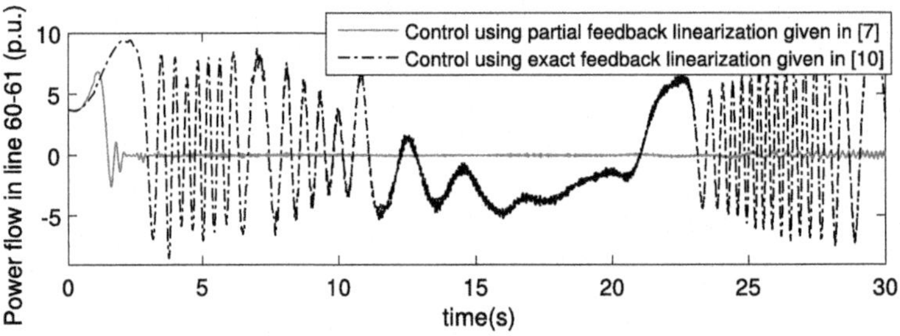

Figure 9.5 Performance of other methods of control based on normal forms.

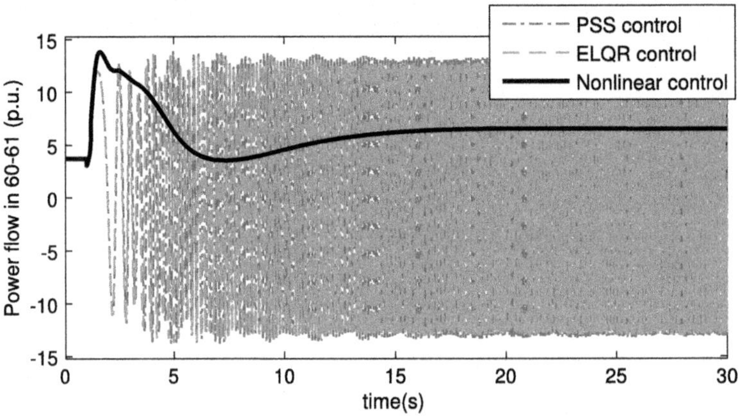

Figure 9.6 Transient stability comparison: Power flow in interarea line.

for the methods based on normal forms are shown in Fig. 9.5). This is because all these tested methods have been derived using classical models of machines and employ gross approximations, and hence they exert adverse effects on the stability of the test system, which is modeled using a detailed subtransient model (IEEE Model 2.2) for machines.

9.5.2 Case B: Assessment of transient stability

In the second case, the system starts from steady state, and at $t = 1$ s a three-phase fault is simulated at bus 54, followed by clearing the fault after 200 ms by the opening of circuit breakers on the line 53–54. Fig. 9.6 shows the response of the three control methods to this large disturbance.

Other nonlinear methods of control have not been used for comparison as they failed to perform properly even for a small disturbance

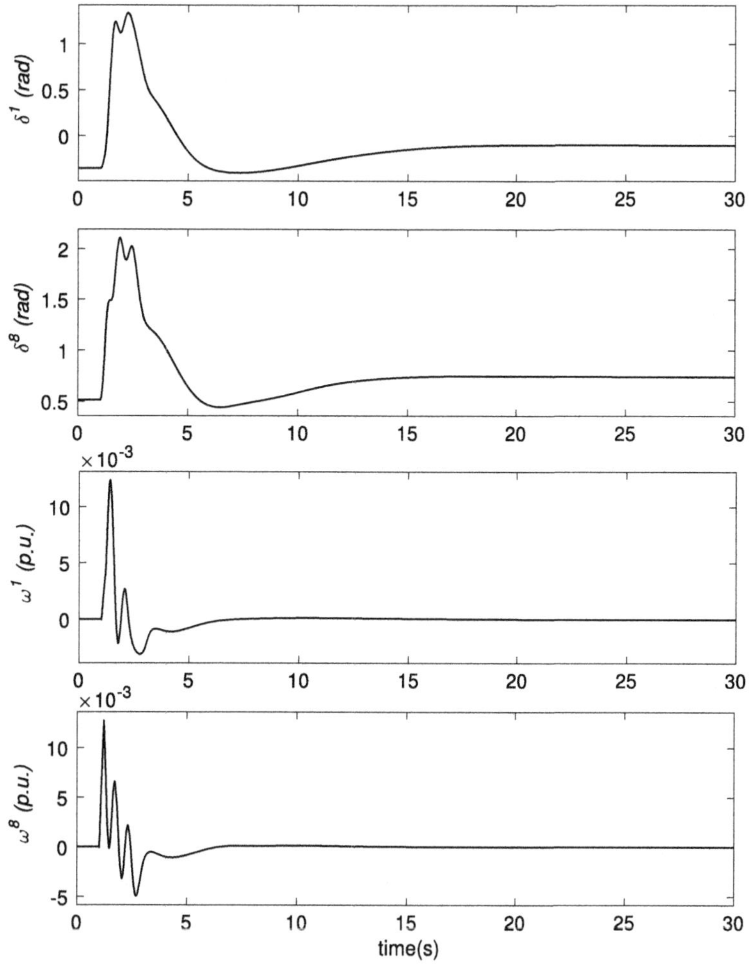

Figure 9.7 States δ and ω for units 1 and 8 under nonlinear control for Case B.

(Remark 9.5). It can be clearly observed that the DSE-NF nonlinear control method can provide adequate damping to the oscillations occurring after the disturbance, and hence they can ensure the transient stability of the system after a large disturbance. On the other hand, the system destabilizes after the disturbance under either PSS control or ELQR control. As both PSS and ELQR fail to stabilize the system, in Figs. 9.7–9.9 only the performance under the DSE-NF nonlinear control method is shown.

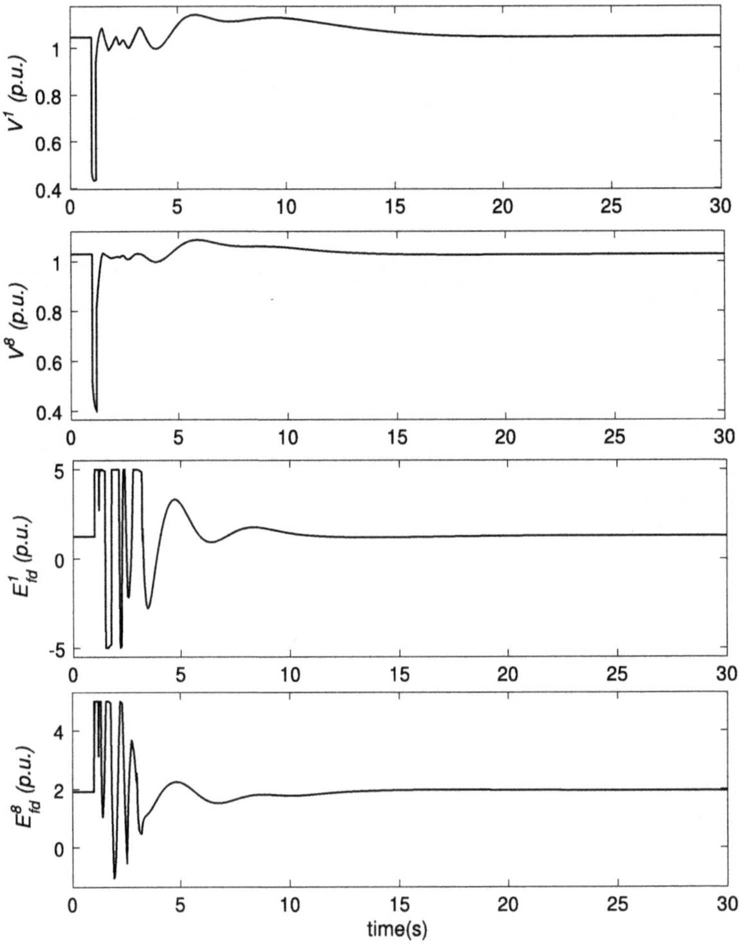

Figure 9.8 V and E_{fd} for units 1 and 8 under nonlinear control for Case B.

9.5.3 Discussion on the magnitude of the control input and the control performance

The final control input that is given to a generator is the excitation voltage, E_{fd}, from the corresponding excitation system. It can be observed in Figs. 9.3, 9.4, 9.8, and 9.9 that the control input is larger for nonlinear control as compared to PSS or ELQR. A relatively large control input is needed in the case of nonlinear control in order to ensure that the control can maintain the stability of the system in face of both large and small disturbances. This observation of nonlinear control requiring relatively large

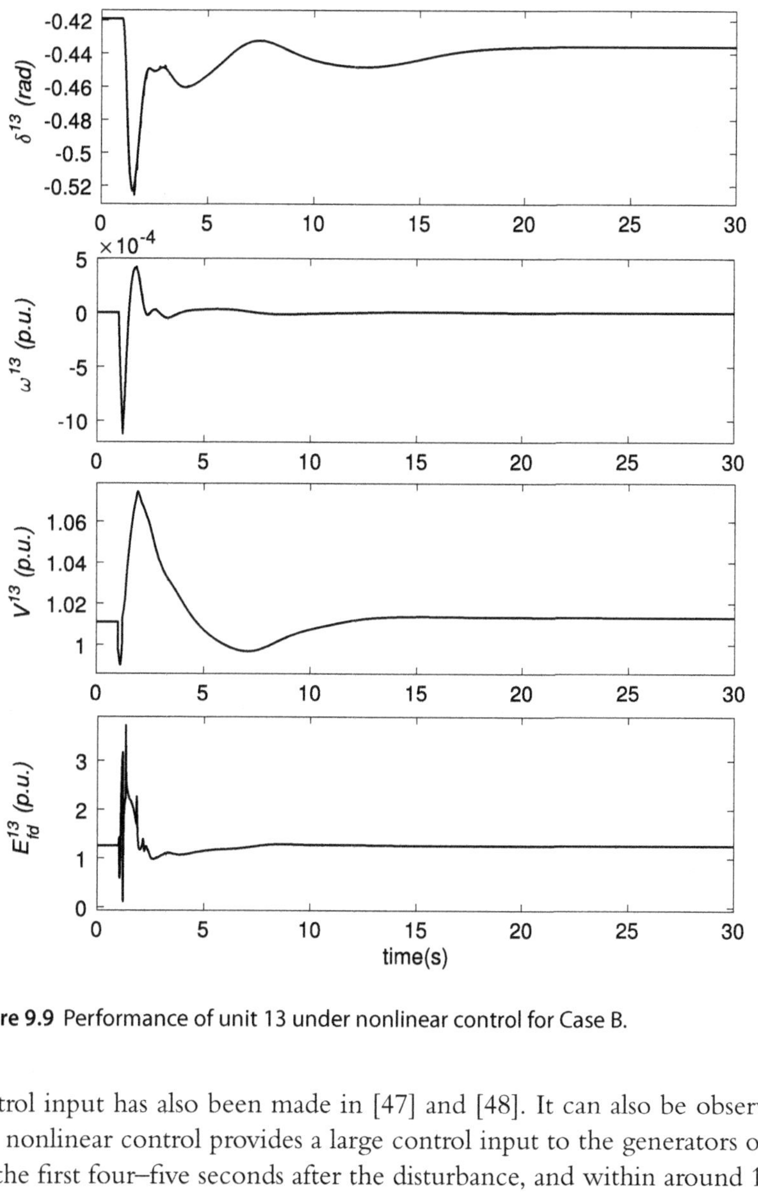

Figure 9.9 Performance of unit 13 under nonlinear control for Case B.

control input has also been made in [47] and [48]. It can also be observed that nonlinear control provides a large control input to the generators only for the first four–five seconds after the disturbance, and within around 15 s the control inputs, the states, and other algebraic variables of generators settle down to their steady-state values. Also, the control input is larger for generators which are electrically close to the disturbance (generators 1 to 8), as opposed to the generators which are not close (such as generator 13).

PSS and ELQR take more time to settle down in the case of a small disturbance (Case A) and fail to maintain the stability of the system in the case of a large disturbance (Case B). This is because the control input from PSS and ELQR is not large enough after the disturbance, and the magnitude of the control input is more or less uniform for all the generators (instead of being larger for generators electrically closer to the disturbance and smaller for machines electrically further), as can be observed in Figs. 9.3, 9.4, 9.8, and 9.9. If larger gains were used for PSS and ELQR, then they would also provide large control inputs. But as PSS and ELQR are derived using a linear model of power systems and are designed specifically for small disturbances, they may not work for large disturbances since a linear model is not valid for large disturbances. Thus, using large gains for PSS and ELQR (or other tools for small signal stability) may worsen the situation in the case of a large disturbance, or jeopardize the stability of the system even in the case of a small disturbance. As the DSE-NF nonlinear control is derived specifically from a nonlinear model of power systems, it is not dependent on a particular equilibrium point or a linear model evaluated at that point. Hence, it ensures that both the small signal stability and the transient stability of the system are maintained irrespective of whether the disturbance is small or large.

9.5.4 Computational feasibility

The complete simulation of the power system, along with the dynamic estimators and the nonlinear controllers at each of the 16 machines, runs in real-time. In the case study, a 30-s simulation takes an average running time of 10.1 s on a personal computer with Intel Core 2 Duo, 2.0 GHz CPU, and 2 GB RAM. Hence the computational requirements at one machine can be easily met for both estimation and control.

9.6 SUMMARY

A nonlinear control scheme has been presented for decentralized control of power system dynamics. This is done by first modeling the subtransient dynamics of power systems in a normal form and then deriving the control law using the normal form. Asymptotic stability of the whole system under the derived control has been proved. It has been demonstrated using simulations that the DSE-NF method can be used to ensure both transient stability and small signal stability. To summarize, the key points described in the chapter are as follows.

- A detailed subtransient model of machines has been used for developing the control method.

- The problem of gross approximation in the final control expression has been eliminated by using estimates of states instead of using directly measured quantities.
- The DSE-NF method is completely decentralized and only requires local measurements at each generation unit.
- It has been rigorously shown that the power system remains asymptotically stable under the developed control.
- The adverse effects of the aforementioned approximations on the stability of the system have been demonstrated.

CHAPTER 10

Conclusion

The importance of optimal control of power systems, so that they operate within their stability margins, is undisputed in today's age of growing power requirements and bulky power system architecture. Having elaborated on the limitations of currently used static control and monitoring tools, reliant on EMSs and steady-state analysis, this book has sought to conduct an in-depth study of the methods for dynamic estimation and control of power systems. Preliminary research in dynamic estimation and control explored centralized schemes, the main drawback of which, the unavailability of a fast, reliable, and secure communication network, has led to the development of alternate decentralized dynamic estimation and control techniques. The limitations of centralized schemes have been highlighted and various decentralized schemes have been discussed in detail. It has been shown that the methods presented are practical for current power systems and these methods have also been demonstrated on a benchmark power system model. Specifically, after reviewing the basic concepts of power system modeling and simulation, this book has:

1. presented the challenges of centralized estimation and control of power systems using currently available communication network and
2. presented different decentralized algorithms for dynamic state estimation and used the estimates thus obtained, in order to generate various decentralized control laws. The integrated control schemes have been demonstrated to be completely decentralized, thereby providing a practical alternative for eliminating the requirement of a fast and reliable communication network necessary for a centralized control approach.

In analyzing the stability effects of introducing a packet-based communication network in the control loops of a power system, this book has presented a generalized framework to assess the effects of packet dropout on the oscillatory stability response of a networked controlled power system (NCPS). A formal approach has also been presented to compute the lower limit on the probability of packet dropout to guarantee the specified damping and stability margins for any operating condition of a power system.

The NCPS model, however, assumes applicability only with update rates of faster communication networks than those used in present-day

Dynamic Estimation and Control of Power Systems
https://doi.org/10.1016/B978-0-12-814005-5.00021-2

power systems. Hence, the book progresses to the next logical examination of performing the dynamic estimation and control of power systems in a decentralized manner, as that bypasses the need of a fast and reliable communication network. First, a scheme for decentralized estimation of the dynamic states of a power system has been explained. The scheme rests on the concept of treating some of the measured signals as pseudoinputs and utilizes unscented Kalman filtering (UKF) for dynamic state estimation of power systems. The feasibility, speed, simplicity, and accuracy of the proposed scheme over centralized schemes has been demonstrated. A comparison of the results obtained clearly establishes the decentralized method to be very advantageous for dynamic state estimation for dynamic control and dynamic security assessment in modern power systems. The problems associated with time synchronization through GPS have also been presented, and a solution to these problems has been presented in the form of a combination of techniques of signal parameter estimation using interpolated discrete Fourier transform and dynamic state estimation using UKF.

Next, the book proceeds to discuss various control laws that use the estimated dynamic states as inputs. A control scheme, termed extended linear quadratic regulator (ELQR), has been presented for the optimal control of a linear system in which both traditional inputs and pseudoinputs are present. Demonstrating the applicability of the control scheme on a simple model linear time-invariant (LTI) system, the control law has been found to significantly minimize the state deviation costs and the required control effort, compared to the classical LQR scheme. The concepts of decentralized dynamic state estimation and ELQR are integrated for the ultimate objective of decentralized estimation and control of a power system. A nonlinear control scheme based on normal forms and detailed power system models has also been presented and combined with decentralized dynamic state estimation for optimal control of power systems, ensuring both transient stability and small signal stability.

All the presented decentralized estimation and control schemes have been implemented and validated on a benchmark 68-bus 16-machine system in MATLAB. The results obtained and comparisons with existing centralized and decentralized control schemes prove that the fully decentralized integrated control scheme not only dispenses with the communication requirements of the centralized control schemes, but is also optimal, computationally feasible, and easily implementable.

APPENDIX A

Description of the 16-Machine, 68-Bus, 5-Area Test System

The 68-bus system (whose line diagram is shown in Fig. A.1) is a reduced-order equivalent of the interconnected New England test system (NETS) (containing G1 to G9) and New York power system (NYPS) (containing G10 to G13), with five geographical regions out of which NETS and NYPS are represented by a group of generators, whereas the power import from each of the three neighboring areas is approximated by equivalent generator models (G14 to G16). There are three major tie lines between NETS and NYPS (connecting buses 60–61, 53–54, and 27–53). All the three are double-circuit tie lines. Generators G1 to G8 have DC excitation systems of type IEEE-DC1A; G9 has fast static excitation of type IEEE-ST1A, while the rest of the generators have manual excitation. G9 is also equipped with a PSS in order to damp a local mode. Data for the system have been extracted from [87], and, with the MATLAB-Simulink simulation code for this system given in [134] (this code does not consider any

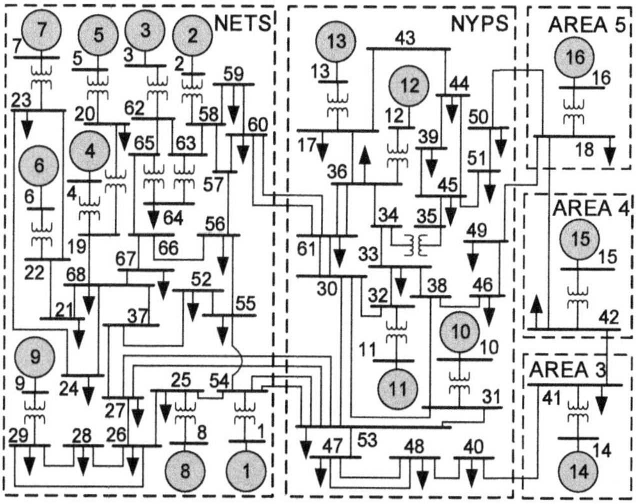

Figure A.1 Line diagram of the 16-machine, 68-bus, 5-area power system.

TCSC in the system and has altered excitation systems and PSSs for the machines), the data are given as follows.

A.1 SYSTEM DATA

A.1.1 Bus data

The base Mega volt–ampere (MVA) for the system is taken as 100 MVA. Table A.1 presents bus data for the system. The bus type in last column has been denoted as 1 for swing buses, 2 for generator buses (PV buses) and 3 for load buses (PQ buses).

Table A.1 Bus data for the 68-bus system

Bus no.	V (p.u.)	P_G (p.u.)	P_L (p.u.)	Q_L (p.u.)	Bus type
1	1.045	2.5	0	0	2
2	0.98	5.45	0	0	2
3	0.983	6.5	0	0	2
4	0.997	6.32	0	0	2
5	1.011	5.05	0	0	2
6	1.05	7	0	0	2
7	1.063	5.6	0	0	2
8	1.03	5.4	0	0	2
9	1.025	8	0	0	2
10	1.01	5	0	0	2
11	1	10	0	0	2
12	1.0156	13.5	0	0	2
13	1.011	35.91	0	0	2
14	1	17.85	0	0	2
15	1	10	0	0	2
16	1	40	0	0	1
17	1	0	60	3	3
18	1	0	24.7	1.23	3
19	1	0	0	0	3
20	1	0	6.8	1.03	3
21	1	0	2.74	1.15	3
22	1	0	0	0	3
23	1	0	2.48	0.85	3
24	1	0	3.09	−0.92	3
25	1	0	2.24	0.47	3

(continued on next page)

Table A.1 (*continued*)

Bus no.	V (p.u.)	P_G (p.u.)	P_L (p.u.)	Q_L (p.u.)	Bus type
26	1	0	1.39	0.17	3
27	1	0	2.81	0.76	3
28	1	0	2.06	0.28	3
29	1	0	2.84	0.27	3
30	1	0	0	0	3
31	1	0	0	0	3
32	1	0	0	0	3
33	1	0	1.12	0	3
34	1	0	0	0	3
35	1	0	0	0	3
36	1	0	1.02	−0.1946	3
37	1	0	0	0	3
38	1	0	0	0	3
39	1	0	2.67	0.126	3
40	1	0	0.6563	0.2353	3
41	1	0	10	2.5	3
42	1	0	11.5	2.5	3
43	1	0	0	0	3
44	1	0	2.6755	0.0484	3
45	1	0	2.08	0.21	3
46	1	0	1.507	0.285	3
47	1	0	2.0312	0.3259	3
48	1	0	2.412	0.022	3
49	1	0	1.64	0.29	3
50	1	0	1	−1.47	3
51	1	0	3.37	−1.22	3
52	1	0	1.58	0.3	3
53	1	0	2.527	1.1856	3
54	1	0	0	0	3
55	1	0	3.22	0.02	3
56	1	0	2	0.736	3
57	1	0	0	0	3
58	1	0	0	0	3
59	1	0	2.34	0.84	3
60	1	0	2.088	0.708	3
61	1	0	1.04	1.25	3
62	1	0	0	0	3
63	1	0	0	0	3
64	1	0	0.09	0.88	3
65	1	0	0	0	3
66	1	0	0	0	3
67	1	0	3.2	1.53	3
68	1	0	3.29	0.32	3

A.1.2 Line data

Table A.2 presents line data for the system.

Table A.2 Line data for the 68-bus system

From bus	To bus	R_L (p.u.)	X_L (p.u.)	Line charging (p.u.)	Tap ratio
1	54	0	0.0181	0	1.025
2	58	0	0.025	0	1.07
3	62	0	0.02	0	1.07
4	19	0.0007	0.0142	0	1.07
5	20	0.0009	0.018	0	1.009
6	22	0	0.0143	0	1.025
7	23	0.0005	0.0272	0	1
8	25	0.0006	0.0232	0	1.025
9	29	0.0008	0.0156	0	1.025
10	31	0	0.026	0	1.04
11	32	0	0.013	0	1.04
12	36	0	0.0075	0	1.04
13	17	0	0.0033	0	1.04
14	41	0	0.0015	0	1
15	42	0	0.0015	0	1
16	18	0	0.003	0	1
17	36	0.0005	0.0045	0.32	1
18	49	0.0076	0.1141	1.16	1
18	50	0.0012	0.0288	2.06	1
19	68	0.0016	0.0195	0.304	1
20	19	0.0007	0.0138	0	1.06
21	68	0.0008	0.0135	0.2548	1
22	21	0.0008	0.014	0.2565	1
23	22	0.0006	0.0096	0.1846	1
24	23	0.0022	0.035	0.361	1
24	68	0.0003	0.0059	0.068	1
25	54	0.007	0.0086	0.146	1
26	25	0.0032	0.0323	0.531	1
27	37	0.0013	0.0173	0.3216	1
27	26	0.0014	0.0147	0.2396	1
28	26	0.0043	0.0474	0.7802	1
29	26	0.0057	0.0625	1.029	1
29	28	0.0014	0.0151	0.249	1

(continued on next page)

Table A.2 *(continued)*

From bus	To bus	R_L (p.u.)	X_L (p.u.)	Line charging (p.u.)	Tap ratio
30	53	0.0008	0.0074	0.48	1
30	61	0.00095	0.00915	0.58	1
31	30	0.0013	0.0187	0.333	1
31	53	0.0016	0.0163	0.25	1
32	30	0.0024	0.0288	0.488	1
33	32	0.0008	0.0099	0.168	1
34	33	0.0011	0.0157	0.202	1
34	35	0.0001	0.0074	0	0.946
36	34	0.0033	0.0111	1.45	1
36	61	0.0011	0.0098	0.68	1
37	68	0.0007	0.0089	0.1342	1
38	31	0.0011	0.0147	0.247	1
38	33	0.0036	0.0444	0.693	1
40	41	0.006	0.084	3.15	1
40	48	0.002	0.022	1.28	1
41	42	0.004	0.06	2.25	1
42	18	0.004	0.06	2.25	1
43	17	0.0005	0.0276	0	1
44	39	0	0.0411	0	1
44	43	0.0001	0.0011	0	1
45	35	0.0007	0.0175	1.39	1
45	39	0	0.0839	0	1
45	44	0.0025	0.073	0	1
46	38	0.0022	0.0284	0.43	1
47	53	0.0013	0.0188	1.31	1
48	47	0.00125	0.0134	0.8	1
49	46	0.0018	0.0274	0.27	1
51	45	0.0004	0.0105	0.72	1
51	50	0.0009	0.0221	1.62	1
52	37	0.0007	0.0082	0.1319	1
52	55	0.0011	0.0133	0.2138	1
54	53	0.0035	0.0411	0.6987	1
55	54	0.0013	0.0151	0.2572	1
56	55	0.0013	0.0213	0.2214	1
57	56	0.0008	0.0128	0.1342	1
58	57	0.0002	0.0026	0.0434	1
59	58	0.0006	0.0092	0.113	1
60	57	0.0008	0.0112	0.1476	1
60	59	0.0004	0.0046	0.078	1
61	60	0.0023	0.0363	0.3804	1
63	58	0.0007	0.0082	0.1389	1

(continued on next page)

Table A.2 (*continued*)

From bus	To bus	R_L (p.u.)	X_L (p.u.)	Line charging (p.u.)	Tap ratio
63	62	0.0004	0.0043	0.0729	1
63	64	0.0016	0.0435	0	1.06
65	62	0.0004	0.0043	0.0729	1
65	64	0.0016	0.0435	0	1.06
66	56	0.0008	0.0129	0.1382	1
66	65	0.0009	0.0101	0.1723	1
67	66	0.0018	0.0217	0.366	1
68	67	0.0009	0.0094	0.171	1
27	53	0.032	0.32	0.41	1

Table A.3 Machine data for the 68-bus system (A)

Machine no.	Bus	Base MVA	X_l (p.u.)	R_a (p.u.)	H (s)	D (p.u.)
1	1	100	0.0125	0	42	4
2	2	100	0.035	0	30.2	9.75
3	3	100	0.0304	0	35.8	10
4	4	100	0.0295	0	28.6	10
5	5	100	0.027	0	26	3
6	6	100	0.0224	0	34.8	10
7	7	100	0.0322	0	26.4	8
8	8	100	0.028	0	24.3	9
9	9	100	0.0298	0	34.5	14
10	10	100	0.0199	0	31	5.56
11	11	100	0.0103	0	28.2	13.6
12	12	100	0.022	0	92.3	13.5
13	13	200	0.003	0	248	33
14	14	100	0.0017	0	300	100
15	15	100	0.0017	0	300	100
16	16	200	0.0041	0	225	50

A.1.3 Machine parameters

Tables A.3, A.4, and A.5 present parameters of the 16 machines in the system.

A.1.4 Excitation system parameters

The IEEE-DC1A type of excitation system has the following parameters:

$$T_r = 0.01 \text{ s}, \quad K_a = 40.0 \text{ p.u.}, \quad T_a = 0.02 \text{ s},$$

Table A.4 Machine data for the 68-bus system (B)

Machine no.	X_d (p.u.)	X'_d (p.u.)	X''_d (p.u.)	T'_{d0} (s)	T''_{d0} (s)
1	0.1	0.031	0.025	10.2	0.05
2	0.295	0.0697	0.05	6.56	0.05
3	0.2495	0.0531	0.045	5.7	0.05
4	0.262	0.0436	0.035	5.69	0.05
5	0.33	0.066	0.05	5.4	0.05
6	0.254	0.05	0.04	7.3	0.05
7	0.295	0.049	0.04	5.66	0.05
8	0.29	0.057	0.045	6.7	0.05
9	0.2106	0.057	0.045	4.79	0.05
10	0.169	0.0457	0.04	9.37	0.05
11	0.128	0.018	0.012	4.1	0.05
12	0.101	0.031	0.025	7.4	0.05
13	0.0296	0.0055	0.004	5.9	0.05
14	0.018	0.00285	0.0023	4.1	0.05
15	0.018	0.00285	0.0023	4.1	0.05
16	0.0356	0.0071	0.0055	7.8	0.05

Table A.5 Machine data for the 68-bus system (C)

Machine no.	X_q (p.u.)	X'_q (p.u.)	X''_q (p.u.)	T'_{q0} (s)	T''_{q0} (s)
1	0.069	0.028	0.025	1.5	0.035
2	0.282	0.06	0.05	1.5	0.035
3	0.237	0.05	0.045	1.5	0.035
4	0.258	0.04	0.035	1.5	0.035
5	0.31	0.06	0.05	0.44	0.035
6	0.241	0.045	0.04	0.4	0.035
7	0.292	0.045	0.04	1.5	0.035
8	0.28	0.05	0.045	0.41	0.035
9	0.205	0.05	0.045	1.96	0.035
10	0.115	0.045	0.04	1.5	0.035
11	0.123	0.015	0.012	1.5	0.035
12	0.095	0.028	0.025	1.5	0.035
13	0.0286	0.005	0.004	1.5	0.035
14	0.0173	0.0025	0.0023	1.5	0.035
15	0.0173	0.0025	0.0023	1.5	0.035
16	0.0334	0.006	0.0055	1.5	0.035

$$K_x = 1.0 \text{ p.u.}, \; T_x = 0.785 \text{ p.u.}, \; A_x = 0.07 \text{ p.u.}, \; B_x = 0.91 \text{ p.u.},$$

$$E_{fdmin} = -10 \text{ p.u.}, \; E_{fdmax} = 10 \text{ p.u.}$$

The IEEE-ST1A type of excitation system has the following parameters:

$$T_r = 0.01 \text{ s}, \; K_a = 200.0 \text{ p.u.}, \; E_{fdmin} = -5 \text{ p.u.}, \; E_{fdmax} = 5 \text{ p.u.}$$

A.1.5 PSS parameters

PSS has the following parameters:

$$K_{pss} = 12 \text{ p.u.}, \; T_w = 10 \text{ p.u.}, \; T_{11} = 0.1 \text{ s}, \; T_{12} = 0.2 \text{ s}, \; T_{21} = 0.1 \text{ s},$$

$$T_{22} = 0.2 \text{ s}, \; V_{ssmin} = -0.05 \text{ p.u.}, \; V_{ssmax} = 0.2 \text{ p.u.}$$

A.1.6 TCSC parameters

TCSC (if present on a line) has the following parameters:

$$K_c = 0.5 \text{ p.u.}, \; K_{cmin} = 0.1 \text{ p.u.}, \; K_{cmax} = 0.8 \text{ p.u.}, \; T_{tcsc} = 0.02 \text{ s}.$$

APPENDIX B

Dynamic State Estimation Plots for Unit 3 and Unit 9

The dynamic state estimation plots for the third and the ninth generating units are as follows (see Figs. B.1–B.12).

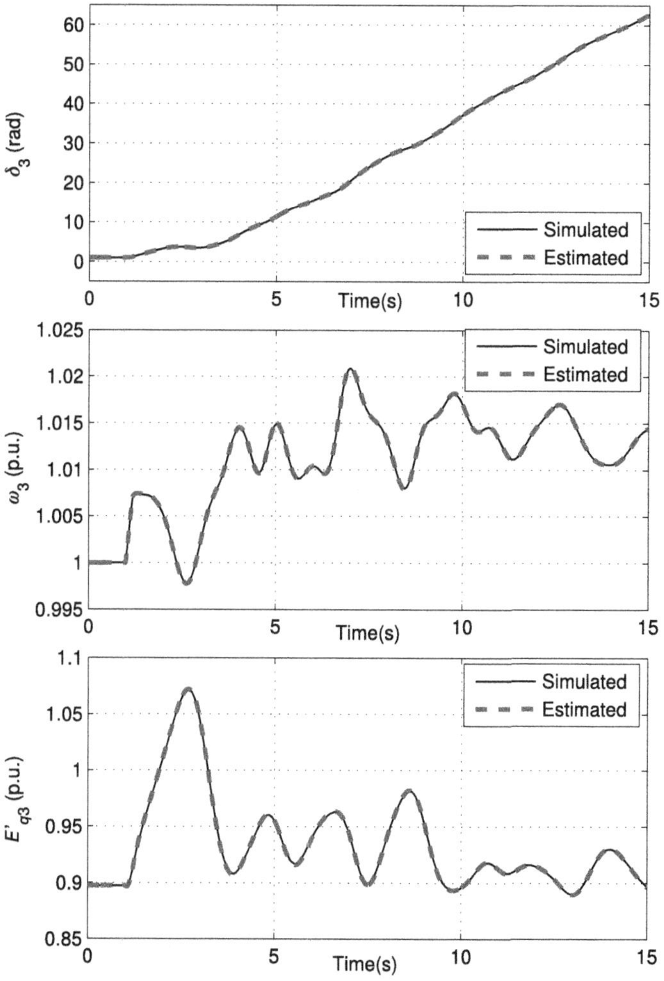

Figure B.1 Estimated vs. simulated values for δ, ω, and E_q' of the third unit.

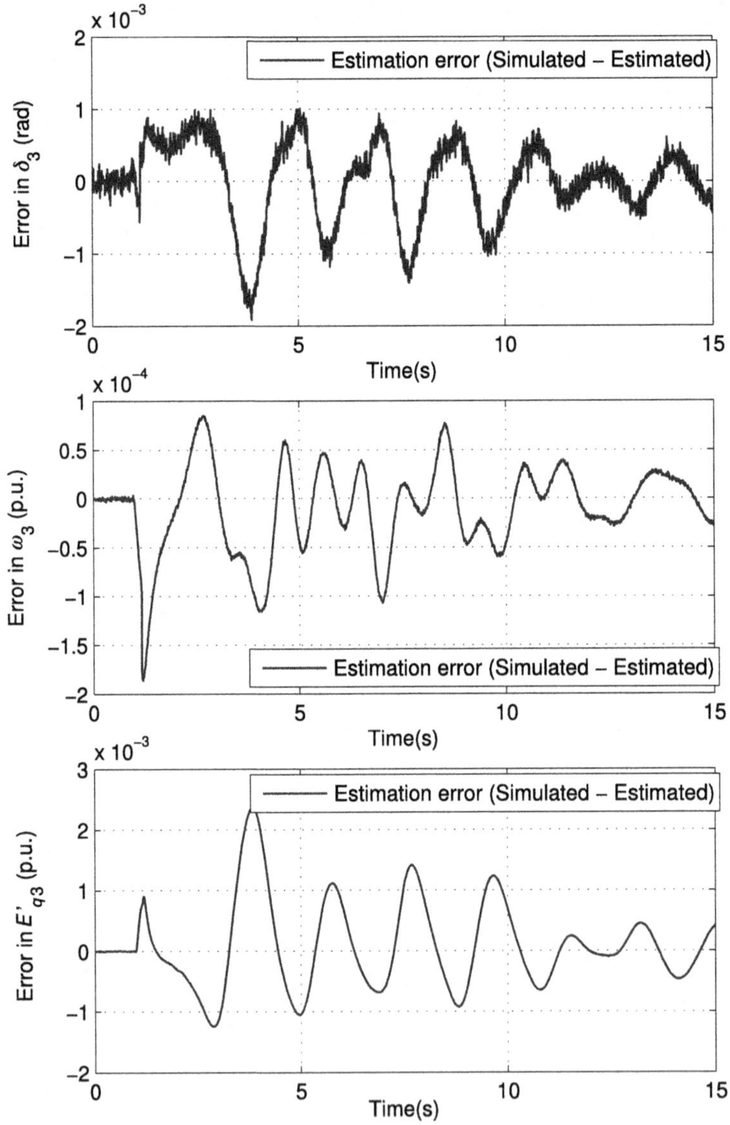

Figure B.2 Estimation errors for δ, ω, and E_q' of the third unit.

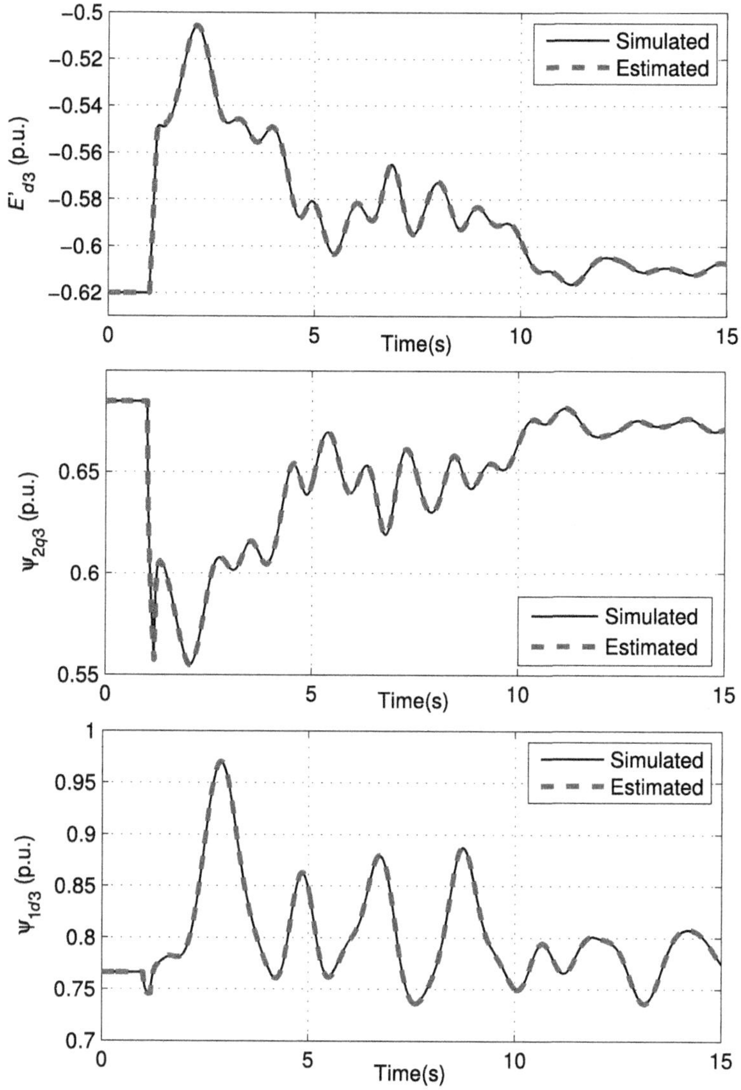

Figure B.3 Estimated vs. simulated values for E'_d, Ψ_{2q}, and Ψ_{1d} of the third unit.

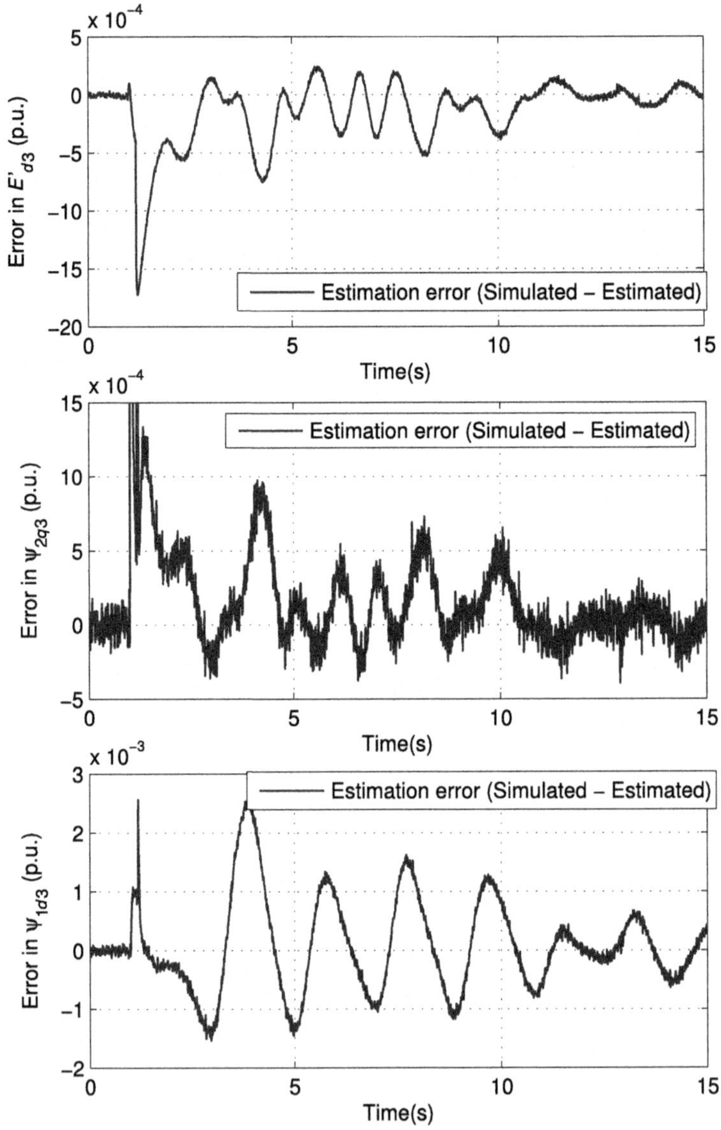

Figure B.4 Estimation errors for E'_d, Ψ_{2q}, and Ψ_{1d} of the third unit.

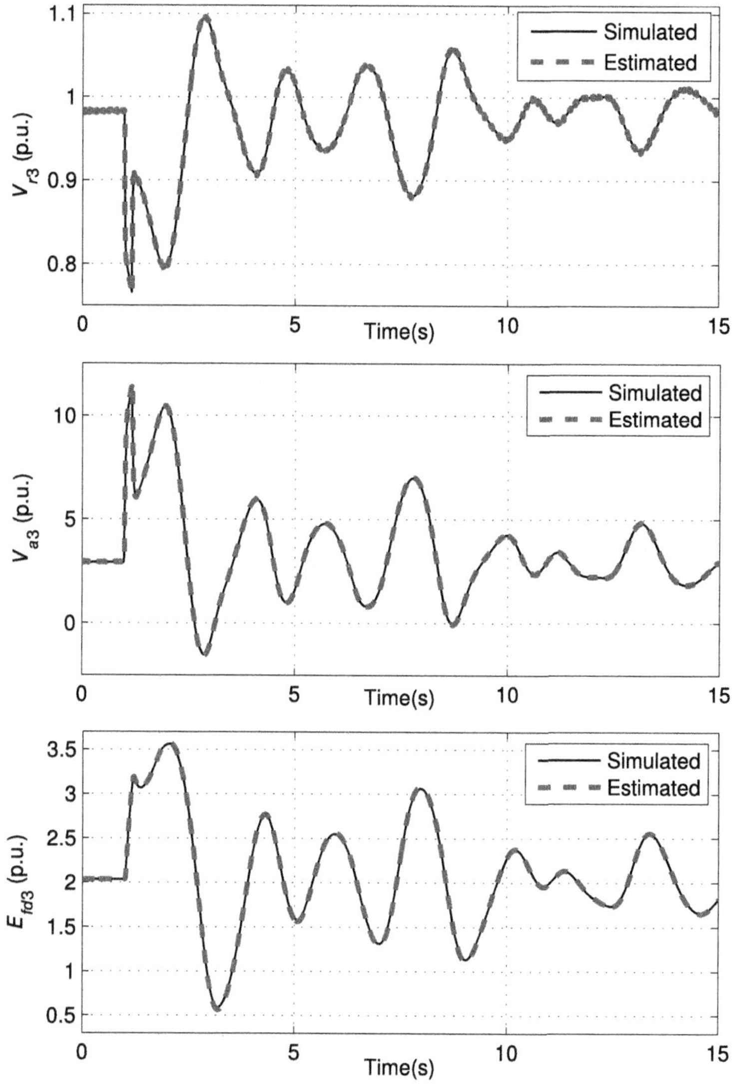

Figure B.5 Estimated vs. simulated values for V_{r3}, V_{a3}, and E_{fd3} of the third unit.

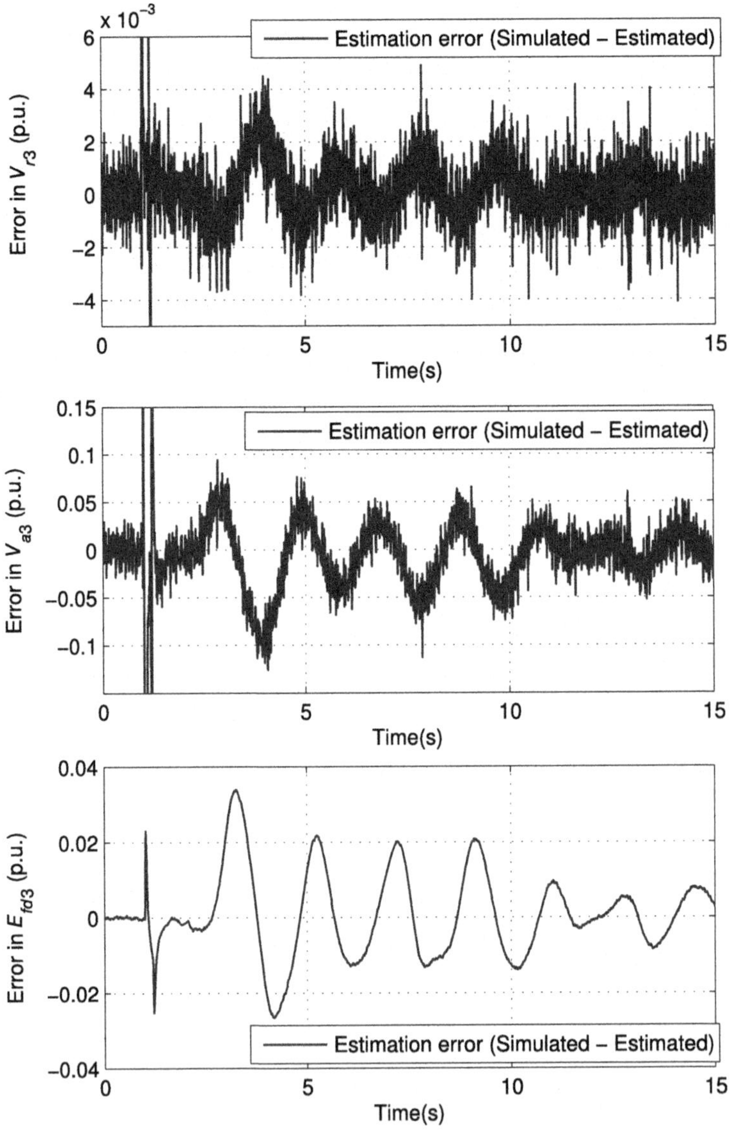

Figure B.6 Estimation errors for V_{r3}, V_{a3}, and E_{fd3} of the third unit.

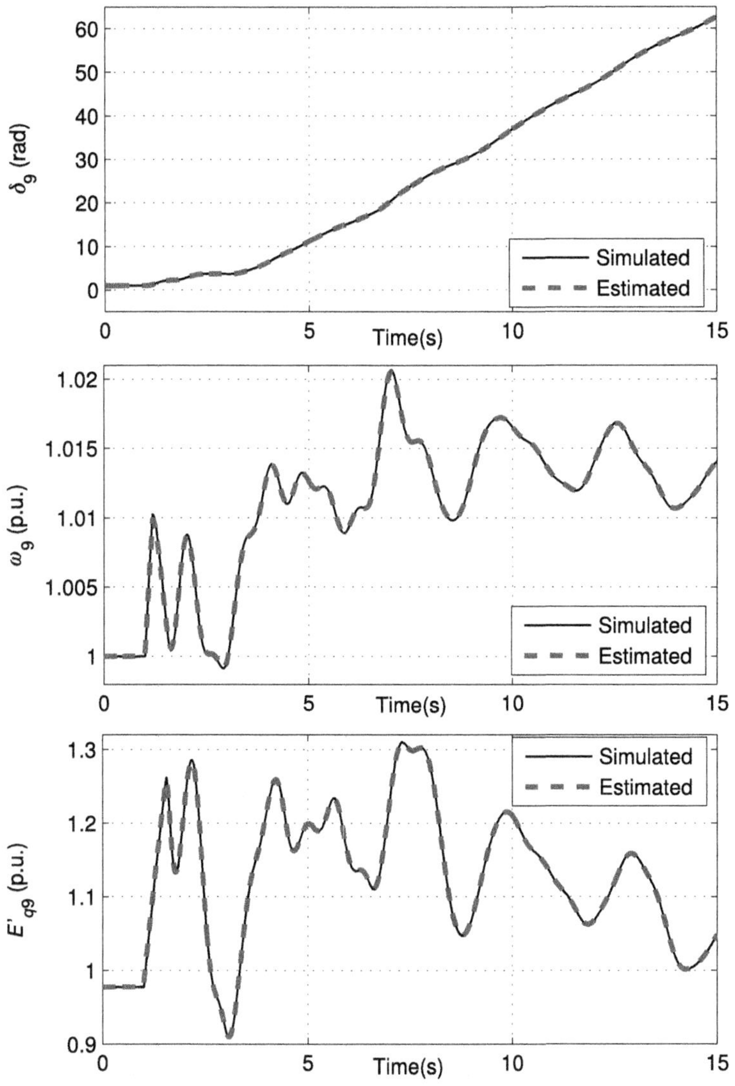

Figure B.7 Estimated vs. simulated values for δ, ω, and E'_q of the ninth unit.

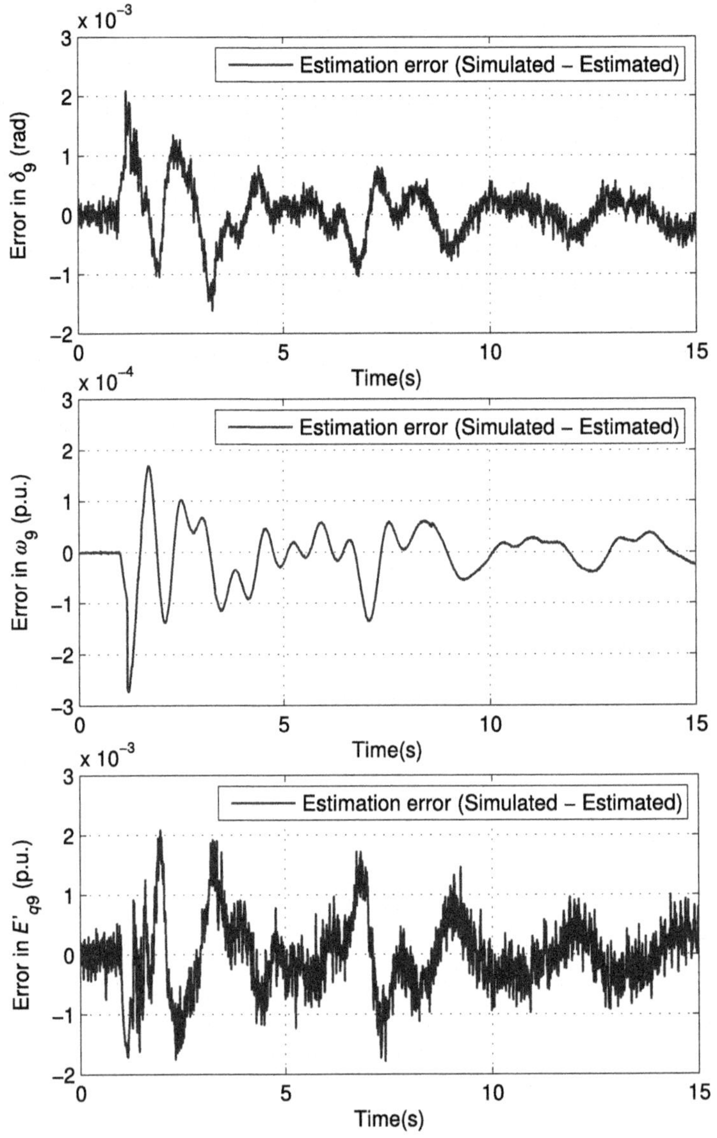

Figure B.8 Estimation errors for δ, ω, and E_q' of the ninth unit.

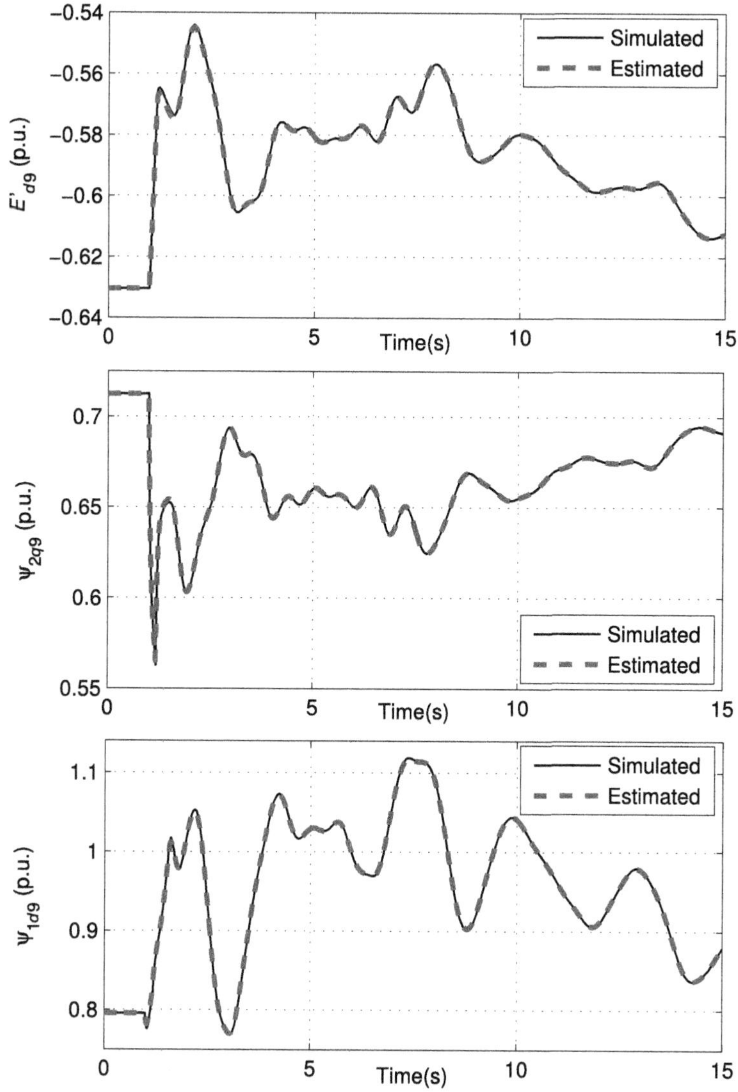

Figure B.9 Estimated vs. simulated values for E'_d, Ψ_{2q}, and Ψ_{1d} of the ninth unit.

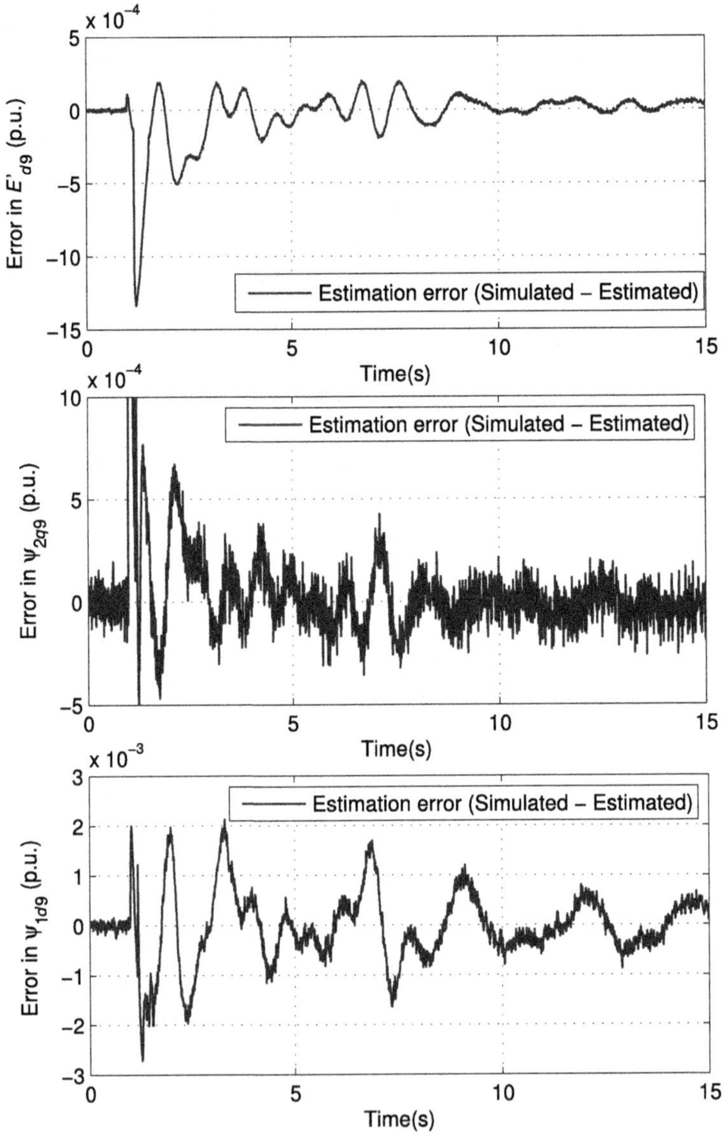

Figure B.10 Estimation errors for E'_d, Ψ_{2q}, and Ψ_{1d} of the ninth unit.

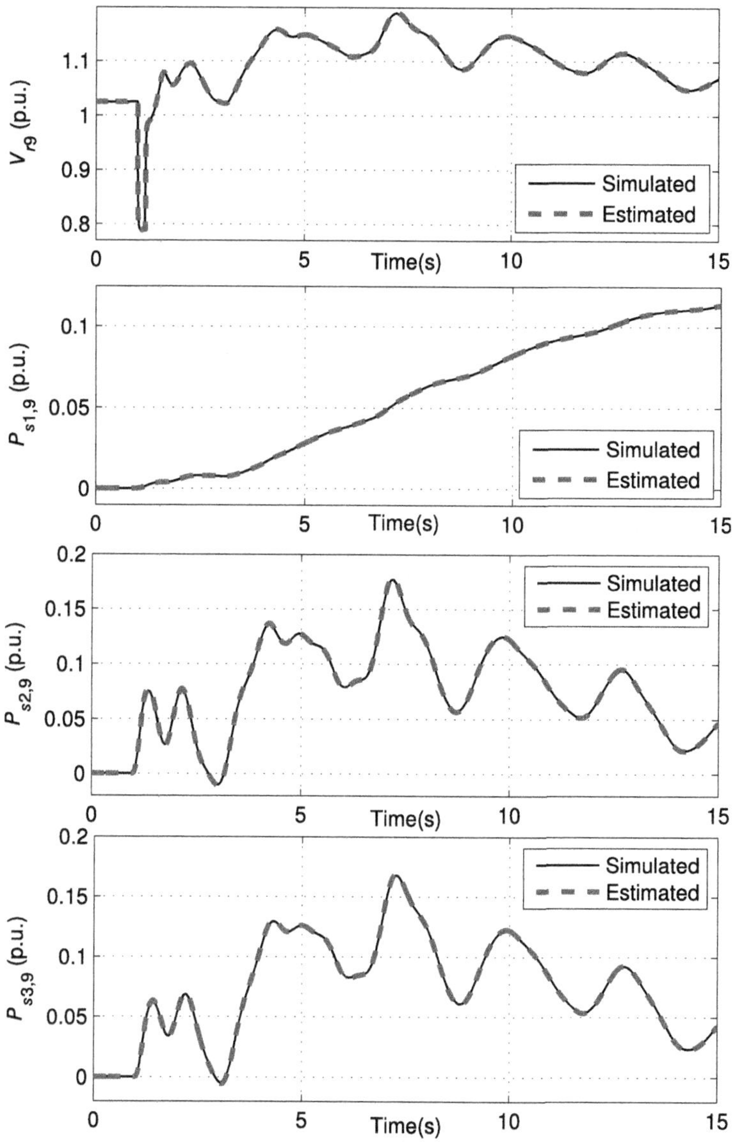

Figure B.11 Estimated vs. simulated values for V_{r9} & PSS states of the ninth unit.

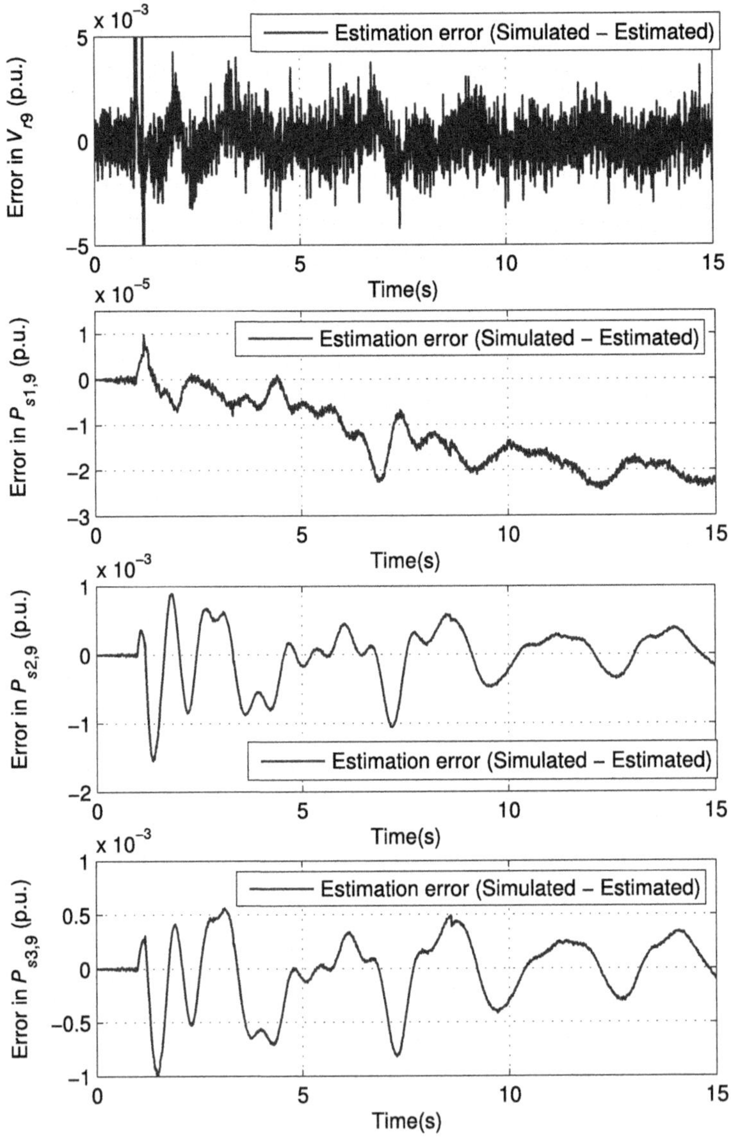

Figure B.12 Estimation errors for V_{r9} & PSS states of the ninth unit.

APPENDIX C

Level-2 S-Function Used in Integrated ELQR

The code for the level-2 S-function used for dynamic updates of system matrices and control law for the integrated ELQR block is as follows.

```
function LQR_Integrated(block)
%Level-2 MATLAB file S-Function for damped ELQR function.
setup(block);
%endfunction

function setup(block)

%% Register number of input and output ports
block.NumInputPorts = 4;%1st input=states, 2nd input=pseudo-inputs, ...
    3rd input=machine parameters, 4th input=sampling period
block.NumOutputPorts = 2;%1st output= state gains, 2nd ...
    output=pseudo-input gains
%% Setup functional port properties to dynamically inherited.
block.SetPreCompInpPortInfoToDynamic;
block.SetPreCompOutPortInfoToDynamic;

block.RegBlockMethod('SetInputPortDimensions', @SetInpPortDims);

block.RegBlockMethod('SetInputPortSamplingMode', @SetInpPortFrameData);

%% Set block sample time to inherited
block.SampleTimes = [-1 0];

%% Set the block simStateCompliance to default (i.e., same as a ...
    built-in block)
block.SimStateCompliance = 'DefaultSimState';

%% Run accelerator on TLC
block.SetAccelRunOnTLC(true);

%% Register methods
block.RegBlockMethod('Outputs',@Output);
%endfunction

function Output(block)

Prm=block.InputPort(3).Data;%machine parameters
T=block.InputPort(4).Data;%the sampling period

xls=   Prm(:,1);
% Ra=    Prm(:,2); %not needed as Ra=0;
xd=    Prm(:,3);
xdd=   Prm(:,4);
xddd=  Prm(:,5);
xq=    Prm(:,6);
xqd=   Prm(:,7);
xqdd=  Prm(:,8);
```

```
Td0d=   Prm(:,9);
Td0dd=  Prm(:,10);
Tq0d=   Prm(:,11);
Tq0dd=  Prm(:,12);
D=      Prm(:,13);
M=      Prm(:,14);
wB=     Prm(:,15);
Ka=     Prm(:,16);
Tr=     Prm(:,17);
kd1=(xddd-xls)./(xdd-xls);
kd2=(xdd-xddd)./(xdd-xls);
kq1=(xqdd-xls)./(xqd-xls);
kq2=(xqd-xqdd)./(xqd-xls);
N_Machine=size(Prm,1);%total no. of machines

x0=block.InputPort(1).Data;%states
alpha0=    x0(:,1);
Ed_dash0= x0(:,3);
Eq_dash0= x0(:,4);
Psi1d0=    x0(:,5);
Psi2q0=    x0(:,6);
%State Edcd is identically zero as xddd=xqdd

u0=block.InputPort(2).Data;%pseudo inputs
Vg=   u0(:,1);
iq0=  (Ed_dash0.*kq1-Psi2q0.*kq2+Vg.*sin(alpha0))./xddd;
id0=  -(Eq_dash0.*kd1+Psi1d0.*kd2-Vg.*cos(alpha0))./xddd;

%%%%%%%%%%%%%partial derivatives%%%%%%%%%%%%%%%%%%%%%%%%
dIq_by_dEdd        =kq1./xddd;
dIq_by_dEqd        =0;
dIq_by_dPsi1d      =0;
dIq_by_dPsi2q      =-kq2./xddd;
dIq_by_dalpha      =Vg.*cos(alpha0)./xddd;
dIq_by_dV          =sin(alpha0)./xddd;

dId_by_dEdd        =0;
dId_by_dEqd        =-kd1./xddd;
dId_by_dPsi1d      =-kd2./xddd;
dId_by_dPsi2q      =0;
dId_by_dalpha      =-Vg.*sin(alpha0)./xddd;
dId_by_dV          =cos(alpha0)./xddd;

dTe_by_dId         =kq1.*Ed_dash0-iq0.*(xddd-xqdd)-Psi2q0.*kq2;
dTe_by_dIq         =kd1.*Eq_dash0-id0.*(xddd-xqdd)+Psi1d0.*kd2;

dTe_by_dEdd         ...
     =id0.*kq1+dTe_by_dId.*dId_by_dEdd+dTe_by_dIq.*dIq_by_dEdd;
dTe_by_dEqd         ...
     =iq0.*kd1+dTe_by_dId.*dId_by_dEqd+dTe_by_dIq.*dIq_by_dEqd;
dTe_by_dPsi1d       ...
     =iq0.*kd2+dTe_by_dId.*dId_by_dPsi1d+dTe_by_dIq.*dIq_by_dPsi1d;
dTe_by_dPsi2q       ...
     =-id0.*kq2+dTe_by_dId.*dId_by_dPsi2q+dTe_by_dIq.*dIq_by_dPsi2q;
dTe_by_dalpha      =dTe_by_dId.*dId_by_dalpha+dTe_by_dIq.*dIq_by_dalpha;
dTe_by_dV          =dTe_by_dId.*dId_by_dV+dTe_by_dIq.*dIq_by_dV;

dfSm_by_dEdd       =-dTe_by_dEdd./M;
dfSm_by_dEqd       =-dTe_by_dEqd./M;
dfSm_by_dPsi1d     =-dTe_by_dPsi1d./M;
dfSm_by_dPsi2q     =-dTe_by_dPsi2q./M;
dfSm_by_dSm        =-D./M;
dfSm_by_dalpha     =-dTe_by_dalpha./M;
dfSm_by_dV         =-dTe_by_dV./M;

```

```
dfEdd_by_dEdd       ...
      =-(1./Tq0d).*(1+(xq-xqd).*(kq1.*dIq_by_dEdd+kq2./(xqd-xls)));
dfEdd_by_dPsi2q     ...
      =-(1./Tq0d).*(xq-xqd).*(kq1.*dIq_by_dPsi2q+kq2./(xqd-xls));
dfEdd_by_dalpha     =-(1./Tq0d).*(xq-xqd).*kq1.*dIq_by_dalpha;
dfEdd_by_dV         =-(1./Tq0d).*(xq-xqd).*kq1.*dIq_by_dV;

dfEqd_by_dEqd       ...
      =-(1./Td0d).*(1+(xd-xdd).*(-kd1.*dId_by_dEqd+kd2./(xdd-xls)));
dfEqd_by_dPsi1d     ...
      =(1./Td0d).*(xd-xdd).*(kd1.*dId_by_dPsi1d+kd2./(xdd-xls));
dfEqd_by_dalpha     =(1./Td0d).*(xd-xdd).*kd1.*dId_by_dalpha;
dfEqd_by_dV         =(1./Td0d).*(xd-xdd).*kd1.*dId_by_dV;
dfEqd_by_dVr        =-Ka./Td0d;

dfPsi1d_by_dEqd     =(1./Td0dd).*(1+(xdd-xls).*dId_by_dEqd);
dfPsi1d_by_dPsi1d   =(1./Td0dd).*(-1+(xdd-xls).*dId_by_dPsi1d);
dfPsi1d_by_dalpha   =(1./Td0dd).*(xdd-xls).*dId_by_dalpha;
dfPsi1d_by_dV       =(1./Td0dd).*(xdd-xls).*dId_by_dV;

dfPsi2q_by_dEdd     =(1./Tq0dd).*(-1+(xqd-xls).*dIq_by_dEdd);
dfPsi2q_by_dPsi2q   =(1./Tq0dd).*(-1+(xqd-xls).*dIq_by_dPsi2q);
dfPsi2q_by_dalpha   =(1./Tq0dd).*(xqd-xls).*dIq_by_dalpha;
dfPsi2q_by_dV       =(1./Tq0dd).*(xqd-xls).*dIq_by_dV;

dfalpha_by_dSm      =wB;

dfVr_by_dVr         =-(1./Tr);
dfVr_by_dV          =(1./Tr);

dEqd_by_dVs         =Ka./Td0d;
%%%%%%%%%%%%partial derivatives end%%%%%%%%%%%%%%%%%

%%%%%%%%%%State matrices' formation using partial derivatives%%%%%%%%%%%%
Kx=zeros(size(x0));
Ku=zeros(size(u0));
N_State=size(x0,2);
Afull =zeros(N_State);
Bfull =zeros(N_State,1);
B1full=zeros(N_State,2);
for i=1:1:N_Machine
    if Ka(i)==0
        continue;
    end

    Afull(3,3)=dfEdd_by_dEdd(i);
    Afull(3,6)=dfEdd_by_dPsi2q(i);
    Afull(3,1)=dfEdd_by_dalpha(i);

    Afull(4,4)=dfEqd_by_dEqd(i);
    Afull(4,5)=dfEqd_by_dPsi1d(i);
    Afull(4,1)=dfEqd_by_dalpha(i);

    Afull(5,4)=dfPsi1d_by_dEqd(i);
    Afull(5,5)=dfPsi1d_by_dPsi1d(i);
    Afull(5,1)=dfPsi1d_by_dalpha(i);

    Afull(6,3)=dfPsi2q_by_dEdd(i);
    Afull(6,6)=dfPsi2q_by_dPsi2q(i);
    Afull(6,1)=dfPsi2q_by_dalpha(i);

    Afull(2,3)=dfSm_by_dEdd(i);
    Afull(2,4)=dfSm_by_dEqd(i);
    Afull(2,5)=dfSm_by_dPsi1d(i);
    Afull(2,6)=dfSm_by_dPsi2q(i);
```

```
        Afull(2,2)=dfSm_by_dSm(i);
        Afull(2,1)=dfSm_by_dalpha(i);

        Afull(1,2)=dfalpha_by_dSm(i);

        Bfull(4,1)=dEqd_by_dVs(i);

        B1full(1,2)=-dfalpha_by_dSm(i);
        B1full(3,1)=dfEdd_by_dV(i);
        B1full(4,1)=dfEqd_by_dV(i);
        B1full(5,1)=dfPsi1d_by_dV(i);
        B1full(6,1)=dfPsi2q_by_dV(i);
        B1full(2,1)=dfSm_by_dV(i);

        if N_State==7
            Afull(4,7)=dfEqd_by_dVr(i);
            Afull(7,7)=dfVr_by_dVr(i);
            B1full(7,1)=dfVr_by_dV(i);
        end
        %%%%%%%%%%State matrices' formation using partial derivatives ends%%%

        %%%%%%%%%%Eigenvalues%%%%%%%%%%%%%%%%%%%%%%%%%%%%%%%%%%%%%
        [¬,lambdar]=eig(Afull,'nobalance');
        lambdar=diag(lambdar);
        omegar=abs(imag(lambdar));
        %%%%%%%%%%Eigenvalues end%%%%%%%%%%%%%%%%%%%%%%%%%%%%%%%%%%%%%%%%%%%%%

        %%%%%%%%%%%%%%%%%%%%%%%Calculation of r and beta%%%%%%%%%%%%%%%%%%%%
        w_max=max(omegar);
        min_damping_ratio=0.15;
        min_damping=cot(acos(min_damping_ratio));
        theta_spiral=w_max*T;%angle of spiral at w
        radius_spiral=exp(-min_damping*theta_spiral);%radius of spiral at w
        spiral_vector=radius_spiral*exp(1i*theta_spiral);

        slope=(sin(theta_spiral)+min_damping*cos(theta_spiral))/...
            (cos(theta_spiral)-min_damping*sin(theta_spiral));
        %slope of perpendicular to tangent of spiral at (R, tht)

        beta=radius_spiral*(cos(theta_spiral)-sin(theta_spiral)/slope);
        %intercept of the line from (R,tht) on x axis
        r=abs(spiral_vector-beta);
        %%%%%%%%%%%%%%%%%%%%%%%Calculation of r and beta ends%%%%%%%%%%%%%%%%

        %%%%%%%%%%Decentralized LQR%%%%%%%%%%%%%%%%%%%%%%%%%%%%%%%%%%%%%%%
        N_State=size(Afull,1);
        csys=ss(Afull,eye(N_State),zeros(1,N_State),0);
        dsys=c2d(csys,T);
        AD=(dsys.a-beta*eye(N_State))/r;
        BD=dsys.b*Bfull/r;
        B1D=dsys.b*B1full/r;
        Q=eye(N_State);
        [PD,¬,KD] = dare(AD,BD,Q,1);
        KVD=KD/(AD-PD\(PD-eye(N_State)))*B1D;
        Kx(i,:)=KD;
        Ku(i,:)=KVD;
    end

    block.OutputPort(1).Data=Kx;
    block.OutputPort(2).Data=Ku;
    %endfunction

    function SetInpPortDims(block, idx, di)

```

```
block.InputPort(idx).Dimensions = di;
if idx==1
    block.OutputPort(1).Dimensions  = di;
end
if idx==2
    block.OutputPort(2).Dimensions  = di;
end
%endfunction

function SetInpPortFrameData(block, idx, fd)

block.InputPort(idx).SamplingMode = fd;
if idx==1
    block.OutputPort(1).SamplingMode  = fd;
    block.OutputPort(2).SamplingMode  = fd;
end
%endfunction
```

BIBLIOGRAPHY

[1] P. Kundur, J. Paserba, V. Ajjarapu, G. Andersson, A. Bose, C. Canizares, N. Hatziargyriou, D. Hill, A. Stankovic, C. Taylor, T. Van Cutsem, V. Vittal, Definition and classification of power system stability: IEEE/CIGRE joint task force on stability terms and definitions, IEEE Trans. Power Syst. 19 (3) (Aug. 2004) 1387–1401.

[2] P. Kundur, Power System Stability and Control, McGraw-Hill Pvt. Limited, India, 1994.

[3] U. Häger, C. Rehtanz, N. Voropai, Monitoring, Control and Protection of Interconnected Power Systems, Springer, Germany, 2014.

[4] P. Kundur, C.W. Taylor, Blackout Experiences and Lessons, Best Practices for System Dynamic Performance, and the Role of New Technologies, IEEE Task Force Report, 2007.

[5] G. Andersson, P. Donalek, R. Farmer, N. Hatziargyriou, I. Kamwa, P. Kundur, N. Martins, J. Paserba, P. Pourbeik, J. Sanchez-Gasca, R. Schulz, A. Stankovic, C. Taylor, V. Vittal, Causes of the 2003 major grid blackouts in North America and Europe, and recommended means to improve system dynamic performance, IEEE Trans. Power Syst. 20 (4) (Nov. 2005) 1922–1928.

[6] D.N. Kosterev, C.W. Taylor, W.A. Mittelstadt, Model validation for the August 10, 1996 WSCC system outage, IEEE Trans. Power Syst. 14 (3) (Aug. 1999) 967–979.

[7] B. Pal, B. Chaudhuri, Robust Control in Power Systems, Springer, New York, USA, 2005.

[8] M. Klein, G.J. Rogers, P. Kundur, A fundamental study of inter-area oscillations in power systems, IEEE Trans. Power Syst. 6 (3) (Aug. 1991) 914–921.

[9] J. Paserba, Analysis and Control of Power System Oscillations, CIGRE special publication 38.01.07, Technical Brochure no. 111, 1996.

[10] E. Vaahedi, Energy management systems, Ch. 10, in: Practical Power System Operation, Wiley, IEEE Press, USA, 2014.

[11] D. Bailey, E. Wright, Practical SCADA for Industry, Newnes, UK, 2003.

[12] K. Barnes, B. Johnson, R. Nickelson, Review of Supervisory Control and Data Acquisition (SCADA) Systems, Report Prepared for the US Department of Energy, Idaho National Engineering and Environmental Laboratory, Jan. 2004.

[13] D.T. Askounis, E. Kalfaoglou, The Greek EMS-SCADA: from the contractor to the user, IEEE Trans. Power Syst. 15 (4) (Nov. 2000) 1423–1427.

[14] C.L. Su, C.N. Lu, M.C. Lin, Wide area network performance study of a distribution management system, Int. J. Electr. Power Energy Syst. 22 (1) (Jan. 2000) 9–14.

[15] F. Maghsoodlou, R. Masiello, T. Ray, Energy management systems, IEEE Power Energy Mag. 2 (5) (Sept.–Oct. 2004) 49–57.

[16] A.G. Phadke, J.S. Thorp, Synchronized Phasor Measurements and Their Applications, Springer, USA, 2008.

[17] A.G. Phadke, Synchronized phasor measurements in power systems, IEEE Comput. Applic. Power 6 (2) (Apr. 1993) 10–15.

[18] IEEE Standard for Synchrophasor Measurements for Power Systems, IEEE Std C37. 118 1-2011, Dec. 2011.

[19] X.P. Zhang, C. Rehtanz, B. Pal, Flexible AC Transmission Systems: Modeling and Control, Springer, Berlin, Germany, 2006.
[20] Y. Zhang, A. Bose, Design of wide-area damping controllers for interarea oscillations, IEEE Trans. Power Syst. 23 (3) (Aug. 2008) 1136–1143.
[21] C.W. Taylor, D.C. Erickson, K.E. Martin, R.E. Wilson, V. Venkatasubramanian, WACS-wide-area stability and voltage control system: R&D and online demonstration, Proc. IEEE 93 (5) (May 2005) 892–906.
[22] C. Hauser, D. Bakken, A. Bose, A failure to communicate: next generation communication requirements, technologies, and architecture for the electric power grid, IEEE Power Energy Mag. 3 (Mar.–Apr. 2005) 47–55.
[23] K. Tomsovic, D. Bakken, V. Venkatasubramanian, A. Bose, Designing the next generation of real-time control, communication, and computations for large power systems, Proc. IEEE 93 (May 2005) 965–979.
[24] Z. Xie, G. Manimaran, V. Vittal, A. Phadke, V. Centeno, An information architecture for future power systems and its reliability analysis, IEEE Trans. Power Syst. 17 (Aug. 2002) 857–863.
[25] E. Ghahremani, I. Kamwa, Dynamic state estimation in power system by applying the extended Kalman filter with unknown inputs to phasor measurements, IEEE Trans. Power Syst. 26 (4) (2011) 2556–2566.
[26] G.K. Gharban, B.J. Cory, Non-linear dynamic power system state estimation, IEEE Trans. Power Syst. 1 (3) (1986) 276–283.
[27] W. Miller, J. Lewis, Dynamic state estimation in power systems, IEEE Trans. Autom. Control 16 (6) (Dec. 1971) 841–846.
[28] E. Ghahremani, I. Kamwa, Online state estimation of a synchronous generator using unscented Kalman filter from phasor measurements units, IEEE Trans. Energy Convers. 26 (4) (2011) 1099–1108.
[29] N. Zhou, D. Meng, S. Lu, Estimation of the dynamic states of synchronous machines using an extended particle filter, IEEE Trans. Power Syst. 28 (4) (Nov. 2013) 4152–4161.
[30] S. Wang, W. Gao, A.P.S. Meliopoulos, An alternative method for power system dynamic state estimation based on unscented transform, IEEE Trans. Power Syst. 27 (2) (May 2012) 942–950.
[31] G. Valverde, V. Terzija, Unscented Kalman filter for power system dynamic state estimation, IET Gener. Transm. Distrib. 5 (1) (Jan. 2011) 29–37.
[32] A. Abur, A. Rouhani, Linear phasor estimator assisted dynamic state estimation, IEEE Trans. Smart Grid 9 (1) (Jan. 2018) 211–219.
[33] J. Zhao, M. Netto, L. Mili, A robust iterated extended Kalman filter for power system dynamic state estimation, IEEE Trans. Power Syst. 32 (4) (July 2017) 3205–3216.
[34] E. Ghahremani, I. Kamwa, Local and wide-area PMU-based decentralized dynamic state estimation in multi-machine power systems, IEEE Trans. Power Syst. 31 (1) (Jan. 2016) 547–562.
[35] Y. Cui, R. Kavasseri, A particle filter for dynamic state estimation in multi-machine systems with detailed models, IEEE Trans. Power Syst. 30 (6) (Nov. 2015) 3377–3385.
[36] K. Emami, T. Fernando, H.H.C. Iu, H. Trinh, K.P. Wong, Particle filter approach to dynamic state estimation of generators in power systems, IEEE Trans. Power Syst. 30 (5) (Sep. 2015) 2665–2675.
[37] J. Qi, K. Sun, W. Kang, Optimal PMU placement for power system dynamic state estimation by using empirical observability gramian, IEEE Trans. Power Syst. 30 (4) (Jul. 2015) 2041–2054.

[38] X. Qing, H.R. Karimi, Y. Niu, X. Wang, Decentralized unscented Kalman filter based on a consensus algorithm for multi-area dynamic state estimation in power systems, Int. J. Electr. Power Energy Syst. 65 (Feb. 2015) 26–33.

[39] N. Zhou, D. Meng, Z. Huang, G. Welch, Dynamic state estimation of a synchronous machine using PMU data: a comparative study, IEEE Trans. Smart Grid 6 (1) (Jan. 2015) 450–460.

[40] F. Aminifar, M. Shahidehpour, M. Fotuhi-Firuzabad, S. Kamalinia, Power system dynamic state estimation with synchronized phasor measurements, IEEE Trans. Instrum. Meas. 63 (2) (Feb. 2014) 352–363.

[41] E. Scholtz, V.D. Donde, J.C. Tournier, Parallel Computation of Dynamic State Estimation for Power System, US Patent Application 13/832,670, filed Mar. 15, 2013.

[42] C. Liu, K. Sun, Z.H. Rather, Z. Chen, C.L. Bak, P. Thogersen, P. Lund, A systematic approach for dynamic security assessment and the corresponding preventive control scheme based on decision trees, IEEE Trans. Power Syst. 29 (2) (Mar. 2014) 717–730.

[43] Y. Xu, Z.Y. Dong, J.H. Zhao, P. Zhang, K.P. Wong, A reliable intelligent system for real-time dynamic security assessment of power systems, IEEE Trans. Power Syst. 27 (3) (Aug. 2012) 1253–1263.

[44] A. Fuchs, M. Imhof, T. Demiray, M. Morari, Stabilization of large power systems using VSC-HVDC and model predictive control, IEEE Trans. Power Deliv. 29 (1) (Feb. 2014) 480–488.

[45] J. Licheng, R. Kumar, N. Elia, Model predictive control-based real-time power system protection schemes, IEEE Trans. Power Syst. 25 (2) (May 2010) 988–998.

[46] A.E. Leon, J.M. Mauricio, J.A. Solsona, Multi-machine power system stability improvement using an observer-based nonlinear controller, Electr. Power Syst. Res. 89 (Aug. 2012) 204–214.

[47] Q. Lu, Y.Z. Sun, S. Mei, Nonlinear Control Systems and Power System Dynamics, Springer, Dordrecht, Netherlands, 2013.

[48] M.A. Mahmud, H.R. Pota, M. Aldeen, M.J. Hossain, Partial feedback linearizing excitation controller for multimachine power systems to improve transient stability, IEEE Trans. Power Syst. 29 (2) (Mar. 2014) 561–571.

[49] W. Yao, L. Jiang, J. Fang, J. Wen, S. Cheng, Decentralized nonlinear optimal predictive excitation control for multi-machine power systems, Int. J. Electr. Power Energy Syst. 55 (2014) 620–627.

[50] Y. Guo, D.J. Hill, Y. Wang, Global transient stability and voltage regulation for power systems, IEEE Trans. Power Syst. 16 (4) (Nov. 2001) 678–688.

[51] Q. Lu, Y.Z. Sun, Z. Xu, T. Mochizuki, Decentralized nonlinear optimal excitation control, IEEE Trans. Power Syst. 11 (4) (Nov. 1996) 1957–1962.

[52] J.W. Chapman, M.D. Ilic, C.A. King, L. Eng, H. Kaufman, Stabilizing a multimachine power system via decentralized feedback linearizing excitation control, IEEE Trans. Power Syst. 8 (3) (Aug. 1993) 830–839.

[53] H. Liu, Z. Hu, Y. Song, Lyapunov-based decentralized excitation control for global asymptotic stability and voltage regulation of multi-machine power systems, IEEE Trans. Power Syst. 27 (4) (Nov. 2012) 2262–2270.

[54] R. Yan, Z.Y. Dong, T.K. Saha, R. Majumder, A power system nonlinear adaptive decentralized controller design, Automatica 46 (2) (Feb. 2010) 330–336.

[55] Y. Wang, D. Cheng, C. Li, Y. Ge, Dissipative Hamiltonian realization and energy-based L2-disturbance attenuation control of multimachine power systems, IEEE Trans. Autom. Control 48 (8) (Aug. 2003) 1428–1433.

[56] Q. Lu, S. Mei, W. Hu, F.F. Wu, Y. Ni, T. Shen, Nonlinear decentralized disturbance attenuation excitation control via new recursive design for multimachine power systems, IEEE Trans. Power Syst. 16 (4) (Nov. 2001) 729–736.

[57] Y. Guo, D.J. Hill, Y. Wang, Nonlinear decentralized control of large-scale power systems, Automatica 36 (9) (Sep. 2000) 1275–1289.

[58] A.C. Zolotas, B. Chaudhuri, I.M. Jaimoukha, P. Korba, A study on LQG/LTR control for damping inter-area oscillations in power systems, IEEE Trans. Control Syst. Technol. 15 (1) (Jan. 2007) 151–160.

[59] H.S. Ko, J. Jatskevich, Power quality control of wind-hybrid power generation system using fuzzy-LQR controller, IEEE Trans. Energy Convers. 22 (2) (Jun. 2007) 516–527.

[60] K.M. Son, J.K. Park, On the robust LQG control of TCSC for damping power system oscillations, IEEE Trans. Power Syst. 15 (4) (Nov. 2000) 1306–1312.

[61] J.C. Seo, T.H. Kim, J.K. Park, S.I. Moon, An LQG based PSS design for controlling the SSR in power systems with series-compensated lines, IEEE Trans. Energy Convers. 11 (2) (Jun. 1996) 423–428.

[62] M.A. Pai, P.W. Sauer, Power System Dynamics and Stability, Prentice Hall, New Jersey, USA, 1998.

[63] K.R. Padiyar, Power System Dynamics: Stability and Control, Anshan Limited, Tunbridge Wells, UK, 2004.

[64] IEEE Recommended Practice for Excitation System Models for Power System Stability Studies, IEEE Std 421.5-2016 (Revision of IEEE Std 421.5-2005), Aug. 26, 2016, pp. 1–207.

[65] M.E. Aboul-Ela, Damping controller design for power system oscillations using global signals, IEEE Trans. Power Syst. 11 (2) (May 1996) 767–773.

[66] F.Y. Wang, D. Liu, Networked Control Systems: Theory and Applications, Springer, London, UK, 2008.

[67] J.E. Flood, Telecommunication Networks, 2nd ed., IET, 1997.

[68] L. Zhang, H. Gao, O. Kaynak, Network-induced constraints in networked control systems – a survey, IEEE Trans. Ind. Inform. 9 (1) (Feb. 2013) 403–416.

[69] M.C.F. Donkers, W.P.M.H. Heemels, N. van de Wouw, L. Hetel, Stability analysis of networked control systems using a switched linear systems approach, IEEE Trans. Autom. Control 56 (9) (Sep. 2011) 2101–2115.

[70] W.P.M.H. Heemels, A.R. Teel, N. van de Wouw, D. Nesic, Networked control systems with communication constraints: trade offs between transmission intervals, delays and performance, IEEE Trans. Autom. Control 55 (8) (Aug. 2010) 1781–1796.

[71] W. Zhang, M. Branicky, S. Phillips, Stability of networked control systems, IEEE Control Syst. 21 (Feb. 2001) 84–99.

[72] S. Wang, X. Meng, T. Chen, Wide-area control of power systems through delayed network communication, IEEE Trans. Control Syst. Technol. 20 (Mar. 2012) 495–503.

[73] H. Li, M.Y. Chow, Z. Sun, Eda-based speed control of a networked dc motor system with time delays and packet losses, IEEE Trans. Ind. Electron. 56 (May 2009) 1727–1735.

[74] N.R. Chaudhuri, D. Chakraborty, B. Chaudhuri, An architecture for FACTS controllers to deal with bandwidth-constrained communication, IEEE Trans. Power Deliv. 1 (Mar. 2011) 188–196.

[75] S.H. Horowitz, A.G. Phadke, Power System Relaying, 3rd ed., John Wiley & Sons Ltd, West Sussex, UK, 2008.

[76] M.G. Chiang, R.Y. Safonov, A Schur method for balanced model reduction, IEEE Trans. Autom. Control AC-34 (1989) 729–733.

[77] K. Ogata, Modern Control Engineering, 4th ed., Prentice Hall PTR, Upper Saddle River, NJ, USA, 2001.

[78] D.P. Bertsekas, Dynamic Programming and Optimal Control, 1st ed., Athena Scientific, Belmont, MA, USA, 1995.

[79] B. Sinopoli, L. Schenato, M. Franceschetti, K. Poolla, S. Sastry, Optimal linear LQG control over Lossy networks without packet acknowledgment, Asian J. Control 10 (Jan. 2008) 3–13.

[80] X. Yu, J.W. Modestino, X. Tian, The accuracy of Markov chain models in predicting packet-loss statistics for a single multiplexer, IEEE Trans. Inf. Theory 54 (1) (Jan. 2008) 489–501.

[81] W.D. Koning, Infinite horizon optimal control of linear discrete time systems with stochastic parameters, Automatica 18 (Apr. 1982) 443–453.

[82] B. Sinopoli, L. Schenato, M. Franceschetti, K. Poolla, M. Jordan, S. Sastry, Kalman filtering with intermittent observations, IEEE Trans. Autom. Control 49 (Sep. 2004) 1453–1464.

[83] L.E. Ghaoui, M.A. Rami, Robust state-feedback stabilization of jump linear systems via LMIs, Int. J. Robust Nonlinear Control 6 (Nov. 1996) 1015–1022.

[84] Y. Loparo, K.A. Fang, Stochastic stability of jump linear systems, IEEE Trans. Autom. Control 47 (Jul. 2002) 1204–1208.

[85] M. Chilali, P. Gahinet, H_∞ design with pole placement constraints: an LMI approach, IEEE Trans. Autom. Control 41 (Mar. 1996) 358–367.

[86] D. Arzelier, J. Bernussou, G. Garcia, Pole assignment of linear uncertain systems in a sector via a Lyapunov-type approach, IEEE Trans. Autom. Control 38 (Jul. 1993) 1128–1132.

[87] G. Rogers, Power System Oscillations, Springer, USA, 2000.

[88] N. Martins, L.T.G. Lima, Determination of suitable locations for power system stabilizers and static VAR compensators for damping electromechanical oscillations in large scale power systems, IEEE Trans. Power Syst. 5 (Nov. 1990) 1455–1469.

[89] L.P. Kunjumuhammed, R. Singh, B.C. Pal, Robust signal selection for damping of inter-area oscillations, IET Gener. Transm. Distrib. 6 (5) (May 2012) 404–416.

[90] A. Heniche, I. Kamwa, Control loops selection to damp inter-area oscillations of electrical networks, IEEE Trans. Power Syst. 17 (2) (May 2002) 378–384.

[91] H. Wang, Selection of robust installing locations and feedback signals of FACTS-based stabilizers in multi-machine power systems, IEEE Trans. Power Syst. 14 (2) (May 1999) 569–574.

[92] M.A.M. Ariff, Adaptive Protection and Control for Wide-Area Blackout Prevention, Ph.D. Thesis, Imperial College London, UK, Jun. 2014.

[93] L. Fan, Y. Wehbe, Extended Kalman filtering based real-time dynamic state and parameter estimation using PMU data, Electr. Power Syst. Res. 103 (Oct. 2013) 168–177.

[94] I. Markovsky, B.D. Moor, Linear dynamic filtering with noisy input and output, Automatica 41 (2005) 167–171.

[95] J.K. Uhlmann, Algorithms for multiple target tracking, Am. Sci. 80 (2) (1992) 128–141.

[96] J.K. Uhlmann, Simultaneous Map Building and Localization for Real-Time Applications, Transfer thesis, University of Oxford, 1994.
[97] S.J. Julier, J.K. Uhlmann, A new extension of the Kalman filter to nonlinear systems, in: The Proceedings of AeroSense: the 11th International Symposium on Aerospace/Defense Sensing, Simulation and Controls, Orlando, Florida, SPIE, 1997.
[98] R.E. Kalman, A new approach to linear filtering and prediction problems, J. Basic Eng. 82 D (1960) 35–45.
[99] S. Julier, J. Uhlmann, H.F. Durrant-Whyte, A new method for the nonlinear transformation of means and covariances in filters and estimators, IEEE Trans. Autom. Control 45 (3) (Mar. 2000) 477–482.
[100] K. Tam, Current-Transformer Phase-Shift Compensation and Calibration, Application Report, Texas Instruments, Literature Number SLAA122, Feb. 2001, pp. 1–6.
[101] IEC Standard for Instrument Transformers, IEC Std 60044 ed. 1.2, 2003.
[102] IEEE Standard Requirements for Instrument Transformers, IEEE Std C57. 13-2008 (Revision of IEEE Std C57 13-1993), Jul. 28, 2008, pp. c1–82.
[103] IEEE Standard for Synchrophasor Measurements for Power Systems, IEEE Std C37. 118 1-2011, Dec. 2011.
[104] IEEE Standard for Synchrophasor Measurements for Power Systems – Amendment 1: Modification of Selected Performance Requirements, IEEE Std C37.118.1a-2014 (Amendment to IEEE Std C37.118.1-2011), Apr. 2014.
[105] S. Kwon, W. Chung, Combined synthesis of state estimator and perturbation observer, ASME J. Dyn. Syst. Meas. Control 125 (1) (Mar. 2003) 19–26.
[106] A. Iserles, A First Course in the Numerical Analysis of Differential Equations, Cambridge University Press, UK, 2008.
[107] T.P. Zieliński, K. Duda, Frequency and damping estimation methods—an overview, Metrol. Meas. Syst. 18 (4) (Jan. 2011) 505–528.
[108] J. Borkowski, D. Kania, J. Mroczka, Interpolated-DFT-based fast and accurate frequency estimation for the control of power, IEEE Trans. Ind. Electron. 61 (12) (Dec. 2014) 7026–7034.
[109] D. Belega, D. Dallet, Frequency estimation via weighted multipoint interpolated DFT, IET Sci. Meas. Technol. 2 (1) (Jan. 2008) 1–8.
[110] P. Handel, Properties of the IEEE-STD-1057 four-parameter sine wave fit algorithm, IEEE Trans. Instrum. Meas. 49 (6) (Dec. 2000) 1189–1193.
[111] S.M. Kay, Fundamentals of Statistical Signal Processing: Estimation Theory, Prentice-Hall, Englewood Cliffs, NJ, 1993.
[112] P. Misra, P. Enge, Global Positioning System: Signals, Measurements and Performance, Ganga-Jamuna Press, Lincoln, MA, 2010.
[113] R.S. Singh, H. Hooshyar, L. Vanfretti, Assessment of time synchronization requirements for phasor measurement units, in: IEEE PowerTech, Eindhoven, Jun. 29–Jul. 2, 2015, pp. 1–6.
[114] H. Kwakernaak, R. Sivan, Linear Optimal Control Systems, Vol. 1, Wiley-Interscience, New York, 1972.
[115] C.D. Johnson, Optimal control of the linear regulator with constant disturbances, IEEE Trans. Autom. Control 13 (4) (Aug. 1968) 416–421.
[116] C.D. Johnson, Accommodation of external disturbances in linear regulator and servomechanism problems, IEEE Trans. Autom. Control 16 (6) (Dec. 1971) 635–644.
[117] K.C. Cheok, N.K. Loh, A ball balancing demonstration of optimal and disturbance-accommodating control, IEEE Control Syst. Mag. 7 (1) (Feb. 1987) 54–57.

[118] E. Menguy, J.L. Boimond, L. Hardouin, J.L. Ferrier, Just-in-time control of timed event graphs: update of reference input, presence of uncontrollable input, IEEE Trans. Autom. Control 45 (11) (Nov. 2000) 2155–2159.
[119] J.M. Yang, S.W. Kwak, Model matching for asynchronous sequential machines with uncontrollable inputs, IEEE Trans. Autom. Control 56 (9) (Sep. 2011) 2140–2145.
[120] J.M. Yang, S.W. Kwak, Corrective control of asynchronous machines with uncontrollable inputs: application to single-event-upset error counters, IET Control Theory Appl. 4 (11) (Nov. 2010) 2454–2462.
[121] Eric Ostertag, Deterministic disturbances compensation. Disturbance observer, in: Mono- and Multivariable Control and Estimation: Linear, Quadratic and LMI Methods, vol. 2, Springer Science & Business Media, Berlin, 2011.
[122] A. Isidori, Nonlinear Control Systems, Springer, London, UK, 1995.
[123] Charles L. Philips, H. Troy Nagel, Digital Control System Analysis and Design, 3rd ed., Prentice Hall, USA, 1994.
[124] K. Furuta, S. Kim, Pole assignment in a specified disk, IEEE Trans. Autom. Control 32 (5) (May 1987) 423–427.
[125] N. Kakimoto, A. Nakanishi, K. Tomiyama, Instability of interarea oscillation mode by autoparametric resonance, IEEE Trans. Power Syst. 19 (4) (Nov. 2004) 1961–1970.
[126] IEEE Guide for Synchronous Generator modelling Practices, Applications in Power System Stability Analyses, IEEE Std 1110-2002 (Revision of IEEE Std 1110-1991), 2003, pp. 1–72.
[127] J.J. Sanchez-Gasca, V. Vittal, M.J. Gibbard, A.R. Messina, D.J. Vowles, S. Liu, U.D. Annakkage, Inclusion of higher order terms for small-signal (modal) analysis: committee report-task force on assessing the need to include higher order terms for small-signal (modal) analysis, IEEE Trans. Power Syst. 20 (4) (Nov. 2005) 1886–1904.
[128] A.K. Singh, B.C. Pal, Decentralized robust dynamic state estimation in power systems using instrument transformers, IEEE Trans. Signal Process. 66 (6) (Mar. 2018) 1541–1550.
[129] A.K. Singh, B.C. Pal, Decentralized nonlinear control for power systems using normal forms and detailed models, IEEE Trans. Power Syst. 33 (2) (Mar. 2018) 1160–1172.
[130] A.K. Singh, B.C. Pal, An extended linear quadratic regulator for LTI systems with exogenous inputs, Automatica 76 (Feb. 2017) 10–16.
[131] A.K. Singh, B.C. Pal, Decentralized control of oscillatory dynamics in power systems using an extended LQR, IEEE Trans. Power Syst. 31 (3) (May 2016) 1715–1728.
[132] A.K. Singh, R. Singh, B.C. Pal, Stability analysis of networked control in smart grids, IEEE Trans. Smart Grid 6 (1) (Jan. 2015) 381–390.
[133] A.K. Singh, B.C. Pal, Decentralized dynamic state estimation in power systems using unscented transformation, IEEE Trans. Power Syst. 29 (2) (Mar. 2014) 794–804.
[134] Benchmark systems for small-signal stability analysis and control, online: http://www.sel.eesc.usp.br/ieee/, https://eioc.pnnl.gov/benchmark/ieeess/index.htm.

INDEX

Printed in the United States
By Bookmasters